Springer Series in Materials Science

Volume 269

The Springer Series in Materials Science covers the complete spectrum of materials physics, including fundamental principles, physical properties, materials theory and design. Recognizing the increasing importance of materials science in future device technologies, the book titles in this series reflect the state-of-the-art in understanding and controlling the structure and properties of all important classes of materials.

More information about this series at http://www.springer.com/series/856

Takashi Matsuoka · Yoshihiro Kangawa
Editors

Epitaxial Growth of III-Nitride Compounds

Computational Approach

 Springer

Editors
Takashi Matsuoka
Institute for Materials Research
Tohoku University
Sendai
Japan

Yoshihiro Kangawa
Research Institute for Applied Mechanics
Kyushu University
Kasuga, Fukuoka
Japan

ISSN 0933-033X   ISSN 2196-2812   (electronic)
Springer Series in Materials Science
ISBN 978-3-030-09542-0   ISBN 978-3-319-76641-6   (eBook)
https://doi.org/10.1007/978-3-319-76641-6

Preface

Ever since the developments of blue light-emitting diodes and laser diodes using epitaxial GaN thin films, III-nitride compounds such as AlN, GaN, and InN have been paid much attention for the use of light emission over a wide range of wavelengths. To improve the device performance of these materials, strict control over the growth conditions and thorough understanding of surface reconstructions and the growth kinetics are essential. In particular, the surface reconstructions and the growth kinetics are crucial for understanding the physics and the chemistry on various technological stages in III-nitride growth.

In this book, we present a unified treatment for the growth mechanisms of epitaxial growth in III-nitride compounds on the basis of state-of-the-art computational approach using ab initio calculations, empirical interatomic potentials, and Monte Carlo simulations. This book is the first attempt to gather together the information of theoretical/computational aspects of the growth of III-nitrides, which is scattered in the scientific literature, into a single comprehensive work. The most fundamental and basic aspects of the crystal growth of III-nitride compounds are presented, along with the underlying scientific principles. We also provide the readers with important theoretical aspects of surface structures and elemental growth processes during the epitaxial growth of III-nitride compounds. The book features advanced discussion of fundamental structural and electronic properties, surface structures, fundamental growth processes, and novel behavior of thin films in III-nitride compounds.

This book will serve as a great practical use to researchers, engineers, and graduate students seeking advanced knowledge of the crystal growth and the application of III-nitride compounds. We hope that the book provides the readers with valuable insight and perspective into this rapidly developing and important field.

Some figures in this book were reproduced from several journals, owing to the kind permission granted by authors and publishers. We would like to express our sincere gratitude and deep appreciation to the following publishers: the American Institute of Physics, Japan Society of Applied Physics, Elsevier Science Publisher B.V., and John Wiley & Sons. Funding from the Japan Society for the

Promotion of Science and the Japan Science and Technology Agency is also greatly appreciated.

We have benefitted from many discussions with colleagues about subjects in this book, especially Prof. Takashi Matsuoka of Institute for Materials Research at Tohoku University, Prof. Tadeusz Suski and Prof. Izabela Gorczyca of Institute of High Pressure Physics, Polish Academy of Sciences, Warsaw, Poland.

Tsu, Japan   Toru Akiyama
Tsu, Japan   Tomonori Ito
Fukuoka, Japan   Yoshihiro Kangawa
Chiba, Japan   Takashi Nakayama
Nagoya, Japan   Kenji Shiraishi

Contents

Contributors

Toru Akiyama Department of Physics Engineering, Mie University, Tsu, Japan

Tomonori Ito Department of Physics Engineering, Mie University, Tsu, Japan

Yoshihiro Kangawa Research Institutes for Applied Mechanics, Kyushu University, Fukuoka, Japan

Takashi Nakayama Department of Physics, Chiba University, Chiba, Japan

Kenji Shiraishi Institute of Materials and Systems for Sustainability, Nagoya University, Nagoya, Japan

Chapter 1
Introduction

Tomonori Ito

Since the successful fabrication of high-quality epitaxial GaN in 1990s, which leads to the development of blue light-emitting diodes and laser diodes, a new frontier in optoelectronics have been opened up. This chapter provides the purpose and outline of the book with original references related to the crystal growth of III-nitride compounds, such as AlN, GaN, and InN. The aim of this chapter is to lay out the role of theoretical works in clarifying the epitaxial growth of III-nitrides and to present an overview of the challenges in the theoretical-computational approach for various issues of III-nitride compounds.

1.1 Purpose of the Book

Blue light-emitting diodes (LEDs) and laser diodes (LDs) have been successfully developed because of the successful fabrication of high-quality epitaxial GaN thin films [1–3] along with improvements in device fabrication techniques in 1990s [4, 5]. III-nitride compounds such as AlN and InN in addition to GaN have been also paid much attention because of their unique suitability for light emission over a wide range of wave lengths that was not previously accessible with solid-state light emitters. To achieve the green lasing, for example, there is an increasing interest in crystal growth of $In_xGa_{1-x}N$ alloy system with an In incorporation of approximately 25% in InGaN/GaN quantum wells [6–8]. Moreover, high Al-content AlGaN have attracted much attention as a key material for devices emitting at deep-ultraviolet wave length [9–12]. To improve the device performance of these materials, strict

T. Ito (✉)
Department of Physics Engineering, Mie University, Tsu, Japan
e-mail: tom@phen.mie-u.ac.jp

© Springer International Publishing AG, part of Springer Nature 2018
T. Matsuoka and Y. Kangawa (eds.), *Epitaxial Growth of III-Nitride Compounds*,
Springer Series in Materials Science 269,
https://doi.org/10.1007/978-3-319-76641-6_1

1

control over the growth conditions and thorough understanding of surface recon-structions are essential. The surface structure determines the morphology, the incorporation of the host and the impurity atoms, and the crystal quality. Therefore, the surface reconstructions and the growth kinetics on them are also crucial for understanding the physics and the chemistry on various technological stages in III-nitride growth. So far a lot of theoretical works have been carried out to investigate the surface structures of semiconductors using ab initio calculation that is a promising tool for clarifying the complicated growth processes because of its ability to calculate electronic structures and total energy [13–16]. Other calculation methods, such as empirical interatomic potential, Monte Carlo (MC) simulation, and thermodynamic approach, have also been devised for use in combination with ab initio calculations to elucidate growth-related problems such as dislocation formation, adatom kinetics, and macroscopic behavior during the crystal growth [17, 18]. In this book, we review the development of these computational methods and discuss a wide range of growth-related problems. One such problem is the surface phase diagram calculations incorporating growth conditions that establish the basis of rigorous investigation of the growth-related properties. We provide an overview of these issues and latest achievements to illustrate the capability of the theoretical-computational approach by comparing experimental results and making precise predictions. We will examine the III-nitride compounds including AlN, GaN, InN, and their alloys that are most applicable to LEDs and LDs.

1.2 Outline of the Book

The relationship among these issues including computational methods and various material properties described in this book is shown in Table 1.1. Pseudopotential methods used in our study are typical examples of computational methods for investigating fundamental properties such as crystal structure, electronic band structure, and adsorption energy in this table. This table reveals that these issues are not independent of each other but are closely interrelated. Therefore, these various issues have to be systematically investigated in considering future prospects in computational materials science for semiconductor epitaxial growth including III-nitride growth. To this end, Chap. 2 briefly reviews various computational methods along with their input parameters to present ab initio calculations as the base of the computational methods, empirical interatomic potentials applicable for use in the simulation of complex systems with a large number of atoms, and MC simulations used for investigating kinetic behavior at finite temperatures.

Chapter 3 deals with prototypical results of fundamental properties in bulk state, such as crystal structure and electronic band structure of III-nitride compounds obtained by the ab initio calculations. Furthermore, miscibility of III-nitride alloy semiconductors, dislocation core structure in III-nitride compounds, and composi-tional inhomogeneity around threading dislocations in of III-nitride alloy semicon-ductors are investigated by empirical interatomic potentials and MC simulations.

Table 1.1 Relationship between computational methods and contents described in this book

Computational methods	Contents
Ab initio calculations	3.1 Crystal structure and structural stability 3.2 Electronic band structure 4.1 Surface phase diagram calculations 4.2 Surface reconstructions on III-nitride semiconductor surfaces 4.3 Hydrogen adsorption on III-nitride semiconductor surfaces 5.2 Surface phase diagram of InN under MOVPE condition 5.3 Growth of InN by pressurized-reactor MOVPE 7.1 Atom kinetics on AlN polar surfaces during MOVPE 7.2 Adatom kinetics on semipolar AlN $(11\bar{2}2)$ surface during MOVPE 7.3 Adatom kinetics on semipolar AlN $(1\bar{1}01)$ and $(1\bar{1}02)$ surfaces during MOVPE 7.4 Adsorption behavior of Al and N atoms on nonpolar 4H-SiC $(11\bar{2}2)$ surface 8.1 Surface polarity inversion 8.2 Electron carrier generation by dislocation 8.3 Schottky barrier at metal/InN interface 9.1 Defects in indium-related nitride semiconductors 9.2 Structural design of AlN/GaN superlattices for deep UV LEDs 10.1 Structure and electronic states of Mg incorporated InN surfaces 10.2 Magnesium incorporation on semipolar GaN $(1\bar{1}01)$ surfaces 10.3 Carbon incorporation on semipolar GaN $(1\bar{1}01)$ surfaces 10.4 Stability of nitrogen incorporated Al_2O_3 surfaces 10.5 Chemical and structural change during nitridation of Al_2O_3 surfaces
Empirical interatomic potentials	3.3 Miscibility of III-nitride alloy semiconductors 3.4 Dislocation core structures 3.5 Compositional inhomogeneity around threading dislocations 6.1 Atomic arrangement in InGaN 6.2 In incorporation in InGaN QWs
Monte Carlo simulations	3.5 Compositional inhomogeneity around threading dislocations 6.1 Atomic arrangement in InGaN 7.1 Atom kinetics on AlN polar surfaces during MOVPE 7.2 Adatom kinetics on semipolar AlN $(11\bar{2}2)$ surface during MOVPE 7.3 Adatom kinetics on semipolar AlN $(1\bar{1}01)$ and $(1\bar{1}02)$ surfaces during MOVPE 7.4 Adsorption behavior of Al and N atoms on nonpolar 4H-SiC $(11\bar{2}2)$ surface 10.5 Chemical and structural change during nitridation of Al_2O_3 surfaces
Thermodynamic analysis	5.2 Surface phase diagram of InN under MOVPE condition 5.3 Growth of InN by pressurized-reactor MOVPE 6.2 In incorporation in InGaN QWs

Surface phase diagrams for III-nitride compounds with various surface orientations, which have been explored extensively by ab initio-based approach incorporating growth temperature and gas pressure, are taken up in Chap. 4. This chapter introduces the methodologies and summarizes surface phase diagrams for III-nitride compounds with/without hydrogen important for metal organic vapor phase epitaxy (MOVPE) with hydrogen and molecular beam epitaxy (MBE) without hydrogen. In Chap. 5, MOVPE growth of InN is illustrated by combining thermodynamic approach incorporating Gibbs free energies and ab initio-based approach. In particular, influences of N/III ratio, growth orientation and total pressure on the growth phenomena such as decomposition of material are discussed in detail. Structural stability of InN during pressurized-reactor MOVPE is also clarified in this chapter. Chapter 6 reviews atomic arrangement in InGaN and In incorporation in InGaN quantum well using empirical interatomic potentials and MC simulations. Here the effect of lattice constraint is crucial to control the composition of coherently grown $In_xGa_{1-x}N$ on $In_yGa_{1-y}N$. Systematic investigations of initial growth processes of III-nitride compounds are given in Chap. 7. Adatom kinetics on AlN polar, nonpolar, and semipolar surfaces during MOVPE growth are shown using MC simulations on the basis of the results of adsorption-desorption, and migration energies obtained by ab initio calculations.

From Chaps. 8–10, various specific topics for III-nitride compounds are presented. Chapter 8 describes polarity inversion and electron carrier generation in III-nitride compounds using ab initio calculations. Microscopic mechanism of the surface-polarity inversion is proposed exemplifying AlN growth. Origin of unintentional electron carriers in InN films and around their interfaces and surfaces is also discussed in terms of edge dislocation, Schottky barrier at metal/InN interfaces, and position of the charge neutrality level of InN. Chapter 9 briefly reviews the availability of ab initio approach to clarify the electronic and optical properties of III-nitrides. Unusual narrow band gap found in InN is discussed in terms of N-vacancy formation and large difference in the covalent radius between In and N. The possibility for achieving high emission efficiency in deep-ultraviolet LEDs is also shown utilizing AlN/GaN superlattices. Novel behaviors related to epitaxial growth of III-nitride compounds is summarized in Chap. 10. Surface phase diagram calculations as functions of temperature and gas pressure are extended to investigate Mg incorporation and C incorporation into GaN surfaces during growth. MC simulations are also applied to N incorporation and initial nitridation processes on Al_2O_3 surfaces in this chapter. Several perspectives for realistic simulations are shown to serve as a guideline for controlling a wide range of properties related to III-nitride growth throughout these chapters in this book.

References

1. H. Amano, N. Sawaki, I. Akasaki, Y. Toyoda, Metalorganic vapor phase epitaxial growth of a high quality GaN film using an AlN buffer layer. Appl. Phys. Lett. **48**, 353 (1986)
2. I. Akasaki, H. Amano, Y. Koide, K. Hiramatsu, N. Sawaki, Effects of AlN buffer layer on crystallographic structure and on electrical and optical properties of GaN and $Ga_{1-x}Al_xN$ $(0 < x \leq 0.4)$ films grown on sapphire substrate by MOVPE. J. Cryst. Growth **98**, 209 (1989)

3. S. Nakamura, GaN Growth using GaN buffer layer. Jpn. J. Appl. Phys. **30**, L1705 (1991)
4. H. Amano, M. Kito, K. Hiramatsu, N. Sawaki, I. Akasaki, P-type conduction in Mg-doped GaN treated with low-energy electron beam irradiation (LEEBI). Jpn. J. Appl. Phys. **28**, L2112 (1989)
5. S. Nakamura, T. Mukai, M. Senoh, Candela-class high-brightness InGaN/AlGaN double-heterostructure blue-light-emitting diodes. Appl. Phys. Lett. **64**, 1687 (1994)
6. S.F. Chichibu, H. Yamaguchi, L. Zhao, M. Kubota, T. Onuma, K. Oakamoto, H. Ohata, Improved characteristics and issues of m-plane InGaN films grown on low defect density m-plane freestanding GaN substrates by metalorganic vapor phase epitaxy. Appl. Phys. Lett. **93**, 151908 (2008)
7. D. Queren, A. Avramescu, G. Brüderl, A. Breidenassel, M. Schillgalies, S. Lutgen, U. Strauß, 500 nm electrically driven InGaN based laser diodes. Appl. Phys. Lett. **94**, 081119 (2009)
8. Y. Enya, Y. Yoshizumi, T. Kyono, K. Akita, M. Ueno, M. Adachi, T. Sumitomo, S. Tokuyama, T. Ikegami, K. Katayama, T. Nakamura, 531 nm green lasing of InGaN based laser diodes on semi-polar 2021 free-standing GaN substrates. Appl. Phys. Express **2**, 082101 (2009)
9. T. Nishida, N. Kobayashi, 346 nm emission from AlGaN multi-quantum-well light emitting diode. Phys. Status Solidi A **176**, 45 (1999)
10. V. Adivarahan, W.H. Sun, A. Chitnis, M. Shatalov, S. Wu, H.P. Maruska, M.A. Khan, 250 nm AlGaN light-emitting diodes. Appl. Phys. Lett. **85**, 2175 (2004)
11. M.A. Khan, M. Shatalov, H.P. Maruska, H.M. Wang, E. Kuokstis, III–nitride UV devices. Jpn. J. Appl. Phys. **44**, 7191 (2005)
12. Y. Taniyasu, M. Kasu, T. Makimoto, An aluminium nitride light-emitting diode with a wavelength of 210 nanometres. Nature (London) **441**, 325 (2006)
13. E. Kaxiras, Y. Bar-Yam, J.D. Joannopoulos, K.C. Pandey, Ab initio theory of polar semiconductor surfaces. II. (2×2) reconstructions and related phase transitions of GaAs $(\overline{111})$. Phys. Rev. B **35**, 9625 (1987)
14. G.X. Qian, R.M. Martin, D.J. Chadi, Stoichiometry and surface reconstruction: An ab initio study of GaAs(100) surfaces. Phys. Rev. Lett. **60**, 1962 (1988)
15. J.E. Northrup, Structure of Si(100)H: dependence on the H chemical potential. Phys. Rev. B **44**, 14149 (1991)
16. A.R. Smith, R.M. Feenstra, D.W. Greve, J. Neugebauer, J.E. Northrup, "Reconstructions of the GaN(0001) surface. Phys. Rev. Lett. **79**, 3934 (1997)
17. T. Ito, K. Shiraishi, A Monte Carlo simulation study for adatom migration and resultant atomic arrangements in $Al_xGa_{1-x}As$ on a GaAs(001) surface. Appl. Surf. Sci. **82–83**, 208 (1994)
18. T. Ito, K. Shiraishi, A Monte Carlo simulation study on the structural change of the GaAs (001) surface during MBE. Surf. Sci. **357–358**, 486 (1996)

Part I
Fundamentals of Computational Approach to Epitaxial Growth of III-Nitride Compounds

Chapter 2
Computational Methods

Tomonori Ito and Toru Akiyama

Computational approach to investigate epitaxial growth of III-nitride compounds is primarily concerned with the numerical computation of electronic structures by ab initio calculations and semi-empirical atomistic techniques. Table 2.1 specifies each of these methods including their advantages and disadvantages. Ab initio calculations stand for a group of methods in which geometric and electronics structures can be calculated using the Schrödinger equation with the values of the fundamental constants and the atomic numbers of the atoms present. Geometric and electronic structures of real materials such as III-Nitride compounds can be quantitatively achieved without any empirical parameter by use of ab initio calculations. This method requires no parameters other than fundamental constants of nature, such as mass and charge of electrons and atoms, atomic number, and atomic volume in a given system. However, the applications of ab initio calculations are currently limited due to large computational efforts. In order for applications to dynamic treatments such as molecular dynamics (MD) method and Mote Carlo (MC) method in the simulations of complex systems with a large number of atoms, empirical interatomic potentials for semiconductors have also been proposed by many authors. In this chapter, we provide an overview of various computational methods to discuss a wide range of issues related to III-Nitride compounds. In Sect. 2.1, we briefly describe ab initio calculations. The calculation methods for empirical interatomic potentials are described in Sect. 2.2. MC method, which is used to clarify various phenomena at finite temperature, is also explained in Sect. 2.3.

T. Ito (✉) · T. Akiyama (✉)
Department of Physics Engineering, Mie University, Tsu, Japan
e-mail: tom@phen.mie-u.ac.jp

T. Akiyama
e-mail: akiyama@phen.mie-u.ac.jp

© Springer International Publishing AG, part of Springer Nature 2018   9
T. Matsuoka and Y. Kangawa (eds.), *Epitaxial Growth of III-Nitride Compounds*,
Springer Series in Materials Science 269,
https://doi.org/10.1007/978-3-319-76641-6_2

Table 2.1 Specifications of computational methods in ab initio calculations and empirical interatomic potentials, and their advantages and disadvantages

Method type	Advantages	Disadvantages	Suitable system
Ab initio calculations (using quantum physics)	• Useful for a broad range of systems • Independent of experimental data • Capable of calculating transition states and excited states	• Computationally expensive	• Small systems (hundreds of atoms) • Systems without available experimental data
Empirical interatomic potentials (using classical physics)	• Computationally least intensive • Fast and useful with limited computer resources	• Applicable only for a limited class of system • Unable to calculate electronic properties • Require experimental data	• Large systems (thousands of atoms) • Dynamical processes • Nonequilibrium processes

2.1 Ab Initio Calculations

In this Section, the fundamental aspects of ab initio calculations, such as density-functional theory (DFT) and its approximations for practical applications, are briefly described. The computational techniques that are widely used in current ab initio calculations, such as plane-wave basis set, calculation models, and pseudopotential approach, are also explained.

2.1.1 Density-Functional Theory

Ab initio calculation methods are computational condensed matters physics methods based on DFT, which is a quantum mechanical approach to solve many body electron systems. The density functional theory, developed by Hohenberg and Kohn [1] and Kohn and Sham, [2] provides a simple method for describing exchange and correlation effects of electrons. Hohenberg and Kohn have proved that the total energy is a unique functional of the electron density. The minimum value of the total-energy functional corresponds to the ground-state energy of the system. The density which yields this minimum value is the exact ground-state density. Kohn and Sham [2] have then shown how to replace the many-body electron problem by an exactly equivalent set of self-consistent one-electron equations.

The Kohn-Sham total-energy functional $E_{tot}[n(r)]$ for the ground state of a system of interacting electrons in the external field $v(\mathbf{r})$ can be written as a functional of charge density $n(\mathbf{r})$ (atomic units are used):

$$E_{tot}[n(r)] = T_s[n(r)] + \frac{1}{2}\int d\mathbf{r}d\mathbf{r}' \frac{n(\mathbf{r})n(\mathbf{r}')}{|\mathbf{r} - \mathbf{r}'|} + \int d\mathbf{r}v(\mathbf{r})n(\mathbf{r}) + E_{xc}[n(r)], \quad (2.1)$$

where $T_s[n(r)]$ is the kinetic energy of the non-interacting electrons, the second term is electronic Coulomb interaction, and $E_{xc}[n(r)]$ is the exchange-correlation energy which includes many-body effects of interacting electrons.

According to the minimum property of the total energy for the ground-state charge density, we can derive the following Euler's equation for the N-electron system:

$$\delta\left\{E_{tot}[n(r)] - \mu \int d\mathbf{r}n(\mathbf{r})\right\} = 0, \quad (2.2)$$

where μ is the Lagrange multiplier in the condition for $n(\mathbf{r})$ given by,

$$\int d\mathbf{r}n(\mathbf{r}) = N. \quad (2.3)$$

Equation (2.2) is satisfied if the following set of Kohn-Sham equations is solved self-consistently:

$$\left\{-\frac{1}{2}\nabla^2 + v_{eff}(\mathbf{r})\right\}\psi_i(\mathbf{r}) = \varepsilon_i\psi_i(\mathbf{r}), \quad (2.4)$$

$$v_{eff}(\mathbf{r}) = v(\mathbf{r}) + V_H(\mathbf{r}) + V_{xc}(\mathbf{r}), \quad (2.5)$$

and

$$n(\mathbf{r}) = \sum_{i=1}^{N}|\psi_i(\mathbf{r})|^2. \quad (2.6)$$

Here, $\{\psi_i(\mathbf{r})\}$ are orthonormal eigenfunctions, $\{\varepsilon_i\}$ are the associated eigenvalues of the Schrödinger-type equation in (2.4), $V_H(\mathbf{r})$ is the Hartree potential written as,

$$V_H(\mathbf{r}) = \int d\mathbf{r}' \frac{n(\mathbf{r}')}{|\mathbf{r} - \mathbf{r}'|}, \quad (2.7)$$

and $V_{xc}(\mathbf{r})$ is the exchange-correlation potential defined by the functional derivative expressed as

$$V_{xc}(\boldsymbol{r}) = \frac{\delta E_{xc}[n(\boldsymbol{r})]}{\delta n(\boldsymbol{r})}. \tag{2.8}$$

In the density functional theory, quantum many-body effects are incorporated in the formal exchange-correlation energy functional $E_{xc}[n(\boldsymbol{r})]$. The Kohn-Sham equations represent a mapping of the interacting many-electron system into a system of non-interacting electrons moving in an effective potential $v_{eff}(\boldsymbol{r})$ caused by all the other electrons. If the exchange-correlation energy functional were known exactly, the exchange-correlation potential which includes the effects of exchange and correlation exactly is in principle obtained by taking the functional derivative with respect to the density, and then the many-body problem of interacting electrons is solved exactly. The Kohn-Sham equation must be solved self-consistently so that the occupied electronic states generate a charge density which produces the electronic potential used to construct the equations. The procedure to obtain iterative solution of the Kohn-Sham equations is shown in Fig. 2.1. The total energy is obtained from the eigenfunctions of the Kohn-Sham equation:

$$
\begin{aligned}
E_{tot}[n(\boldsymbol{r})] = {}&-\frac{1}{2}\sum_{i=1}^{N}\int d\boldsymbol{r}\psi_i^*(\boldsymbol{r})\nabla^2\psi_i(\boldsymbol{r}) + \frac{1}{2}\int d\boldsymbol{r}d\boldsymbol{r}'\frac{n(\boldsymbol{r})n(\boldsymbol{r}')}{|\boldsymbol{r}-\boldsymbol{r}'|} \\
&+ \int d\boldsymbol{r}v(\boldsymbol{r})n(\boldsymbol{r}) + E_{xc}[n(\boldsymbol{r})],
\end{aligned} \tag{2.9}
$$

where the density $n(\boldsymbol{r})$ is calculated by using (2.6). In the density functional theory, the total energy is primarily meaningful. The Kohn-Sham eigenvalues are not, strictly speaking, the energies of the quasi-particle electron states, but rather the derivatives of the total energy with respect to the occupation numbers of these

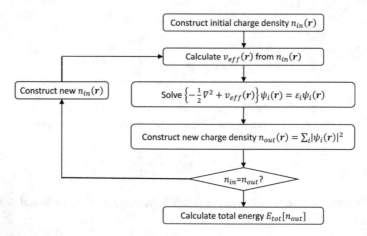

Fig. 2.1 Schematic representation of iterative solution of Kohn-Sham equations. This kind of computation is performed using numerical programs

states. However, in practice, optically observed excitation spectrum is qualitatively understood by the one-electron Kohn-Sham eigenvalues which are solved by (2.4), (2.5), and (2.6).

The simplest method for describing the exchange-correlation energy as a function of the electron density is to use the local density approximation (LDA). In the LDA, the exchange-correlation energy of an electronic system is constructed by assuming that the exchange-correlation energy per electron at a point r is equal to that in a homogeneous electron gas with the same density $n_0 = n(r)$. Therefore, the exchange-correlation energy $E_{xc}[n(r)]$ and the exchange-correlation potential $V_{xc}[n(r)]$ are expressed as

$$E_{xc}[n(r)] = \int dr n(r) \epsilon_{xc}[n(r)], \tag{2.10}$$

and

$$V_{xc}(r) = \frac{\delta E_{xc}[n(r)]}{\delta n(r)} = \frac{\partial [n(r) \epsilon_{xc}[n(r)]]}{\partial n(r)}, \tag{2.11}$$

respectively. Here, $\epsilon_{xc}[n_0]$ is the exchange-correlation energy per electron of the homogeneous system with charge density n_0. For the LDA, the parameterization by Perdew and Zunger [3] which links to the exact exchange-correlation energy of the homogeneous electron gas calculated by Cerpley and Alder using Green's function quantum Monte Carlo method [4] is widely used. The exchange-correlation energy functional of Perdew-Zunger parameterization (including the spin polarization) $E_{xc}[n_\uparrow(r), n_\downarrow(r)]$ in (2.10) is divided to exchange functional $E_x[n_\uparrow(r), n_\downarrow(r)]$ and correlation functional $E_c[n_\uparrow(r), n_\downarrow(r)]$:

$$E_{xc}[n_\uparrow(r), n_\downarrow(r)] = E_x[n_\uparrow(r), n_\downarrow(r)] + E_c[n_\uparrow(r), n_\downarrow(r)], \tag{2.12}$$

For the exchange functional $E_x[n_\uparrow(r), n_\downarrow(r)]$, it is given by

$$E_x[n_\uparrow(r), n_\downarrow(r)] = \frac{1}{2} E_x[n_\uparrow(r)] + \frac{1}{2} E_x[n_\downarrow(r)], \tag{2.13}$$

$$E_x[n(r)] = \int dr n(r) \epsilon_x^{unif}[n(r)], \tag{2.14}$$

where

$$\epsilon_x^{unif}[n(r)] = -\frac{3k_F}{4\pi}, \tag{2.15}$$

$$k_F = \{3\pi^2 n(r)\}^{\frac{1}{3}}. \tag{2.16}$$

For the correlation functional, it is written as

$$E_c\left[n_\uparrow(\boldsymbol{r}), n_\downarrow(\boldsymbol{r})\right] = \int d\boldsymbol{r} n(\boldsymbol{r}) \epsilon_c[r_s(\boldsymbol{r}), \varsigma(\boldsymbol{r})] \tag{2.17}$$

$$r_s(\boldsymbol{r}) = \left(\frac{3}{4\pi n(\boldsymbol{r})}\right)^{\frac{1}{3}}, \tag{2.18}$$

$$\varsigma(\boldsymbol{r}) = \frac{n_\uparrow(\boldsymbol{r}) - n_\downarrow(\boldsymbol{r})}{n(\boldsymbol{r})}, \tag{2.19}$$

$$\epsilon_c[r_s(\boldsymbol{r}), \varsigma(\boldsymbol{r})] = \epsilon_c(r_s, 0) + f[\varsigma(\boldsymbol{r})][\epsilon_c(r_s, 0) - \epsilon_c(r_s, 1)], \tag{2.20}$$

$$f[\varsigma(\boldsymbol{r})] = \frac{\{1 + \varsigma(\boldsymbol{r})\}^{\frac{4}{3}} + \{1 - \varsigma(\boldsymbol{r})\}^{\frac{4}{3}} - 2}{2^{\frac{4}{3}} - 2}, \tag{2.21}$$

where $\epsilon_c(r_s, 0)$ and $\epsilon_c(r_s, 0)$ are fitted by the following functional form $G(r_s)$, which is given by

$$G(r_s) = \begin{cases} \gamma/\left(1 + \beta_1 r_s^{1/2} + \beta_2 r_s\right) & (r_s \geq 1) \\ A \ln r_s + B + C r_s \ln r_s + D r_s & (r_s < 1). \end{cases} \tag{2.22}$$

The fitting parameters are shown in Table 2.2.

In order to solve the Kohn-Sham equation expressed as (2.4), we have to derive the exchange-correlation potential $V_{xc}(\boldsymbol{r}) = \delta E_{xc}[n(\boldsymbol{r})]/\delta n(\boldsymbol{r})$ in (2.8). The functional derivative for the exchange potential for each spin component ($\sigma = \uparrow$ or $\downarrow$) in (2.8) is expressed as,

$$\frac{\delta E_x\left[n_\uparrow(\boldsymbol{r}), n_\downarrow(\boldsymbol{r})\right]}{\delta n_\sigma(\boldsymbol{r})} = \frac{4}{3} \epsilon_x^{unif}[n(\boldsymbol{r})], \tag{2.23}$$

and that for the correlation functional is written as,

Table 2.2 Parameters for the correlation functional form in the LDA by Perdew and Zunger in [3]

	$\epsilon_c(r_s, 0)$	$\epsilon_c(r_s, 1)$
γ	−0.1423	−0.0843
β_1	1.0529	1.3981
β_2	0.3334	0.2611
A	0.0311	0.01555
B	−0.048	−0.0269
C	0.0020	0.0007
D	−0.0116	−0.0048

$$\frac{\delta E_c[n_\uparrow(r), n_\downarrow(r)]}{\delta n_\sigma(r)} = \epsilon_c[r_s(r), \varsigma(r)]$$
$$- \frac{r_s}{3}\frac{\partial \epsilon_c[r_s(r), \varsigma(r)]}{\partial r_s(r)} + (sgn(\sigma) - \varsigma(r))\frac{\partial \epsilon_c[r_s(r), \varsigma(r)]}{\partial \varsigma(r)},$$

$$(2.24)$$

where

$$\frac{\partial \epsilon_c[r_s(r), \varsigma(r)]}{\partial r_s(r)} = \frac{\partial \epsilon_c[r_s(r), 0]}{\partial r_s(r)} + f[\varsigma(r)]\left\{\frac{\partial \epsilon_c[r_s(r), 0]}{\partial r_s(r)} - \frac{\partial \epsilon_c[r_s(r), 1]}{\partial r_s(r)}\right\}, \quad (2.25)$$

$$\frac{\partial \epsilon_c[r_s(r), \varsigma(r)]}{\partial \varsigma(r)} = f'[\varsigma(r)]\{\epsilon_c(r_s, 0) - \epsilon_c(r_s, 1)\}, \tag{2.26}$$

$$f'[\varsigma(r)] = \frac{4}{3}\frac{\{1 + \varsigma(r)\}^{\frac{1}{3}} + \{1 - \varsigma(r)\}^{\frac{1}{3}}}{2^{4/3} - 2}. \tag{2.27}$$

The correlation potential $\partial \epsilon_c[r_s(r), 0]/\partial r_s(r)$ and $\partial \epsilon_c[r_s(r), 1]/\partial r_s(r)$ are also given by the functional from $G(r_s)$, which is given by

$$\frac{dG}{dr_s}(r_s) = \begin{cases} -\gamma\left(\frac{1}{2}\beta_1 r_s^{-\frac{1}{2}} + \beta_2\right)/\left(1 + \beta_1 r_s^{1/2} + \beta_2 r_s\right) & (r_s \geq 1) \\ \frac{A}{r_s} + B + C(\ln r_s + 1) + D & (r_s < 1). \end{cases} \tag{2.28}$$

Detailed description of correlation functionals is discussed in [3].

The LDA has been used successfully in solid state physics: The ground-state properties of atoms, molecules, and solids agree with experiments. Recently, functional forms for the exchange-correlation energy using additional information about the electron gas of slowly varying density have been developed by many authors. In these functional forms called generalized gradient approximation, GGA, the effects of the charge gradient are included. The exchange-correlation functional $E_{xc}[n(r)]$ in the GGA is with the form written as

$$E_{xc}[n(r)] = \int dr n(r) f(n(r), \nabla n(r)). \tag{2.29}$$

This semi-local functional has demonstrated useful improvement over the LDA: In comparison with the LDA, the GGA tend to improve the total energies, atomization energies, and energy barriers. The functional form parameterized by Perdew and Wang (GGA-PW91), [5] and the form by Perdew, Burke, and Ernzerhof (GGA-PBE96) are currently used for solid phase materials [6]. These functionals are free from empirical parameters. Details of the functional forms in the GGA are described in [5, 6].

More recently, a class of approximations to the exchange-correlation energy functional that incorporate a portion of exact exchange energy from Hartree-Fock approximation has been developed as a hybrid density functional approach. The hybrid exchange-correlation functionals are usually constructed as a linear combination of the Hartree–Fock exact exchange functional and exchange and correlation explicit density functionals. The parameters determining the weight of each individual functional are typically determined to reproduce the experimental or accurately calculated thermochemical data. One of the most commonly used versions is B3LYP, which stands for Becke, 3-parameter, and Lee-Yang-Parr [7, 8]. In order to improve computational efficiency, Heyd, Scuseria, and Ernzerhof used an error function screened Coulomb potential to calculate the exchange portion of the energy [9].

The total energy calculations within the LDA and GGA have been successfully used to predict physical properties of solids, such as equilibrium lattice constants, bulk moduli, piezoelectric constants, and phase-transition pressures and temperatures [10, 11]. However, it is well known that the energy gap of semiconductors calculated using the LDA and GGA underestimates the experimentally observed values. One of possible origins for the underestimation of the energy gap is that the density-functional theory in principles guarantees only the ground-state energy of the system. The excited-state energies of semiconductors from the Kohn-Sham equation are not justified. The underestimation can be improved by the calculations of self-energy in terms of the single particle Green's function G and the screened Coulomb interaction W, called GW approximation [12–14].

2.1.2 Plane-Wave Basis Set

Due to the Kohn-Sham equation presented in Sect. 2.1.1, the many-body problem of electrons can be mapped into an effective single-particle problem. However, there still remains a formidable task of handling an infinite number of electrons moving in the infinite number of nuclei or ions. This task can be executed by performing calculations on periodic systems and applying Bloch's theorem to the electronic wavefunctions.

The Bloch's theorem states that in a periodic solid each electronic wavefunction can be written as the product of a cell-periodic part and a wave-like part. The cell-periodic part of the wavefunction can be expanded using a basis set consisting of a discrete set of plane waves whose wave vectors are reciprocal lattice vectors of the cell. Then the wave function has the form written as,

$$\psi_i(r) = \sum_{\mathbf{G}} c_{i,\mathbf{k}+\mathbf{G}} \exp\{i(\mathbf{k}+\mathbf{G}) \cdot r\}, \qquad (2.30)$$

Here the reciprocal lattice vectors $\mathbf{G}$ are defined by $\mathbf{G} \cdot l = 2\pi m$ (l is the lattice vector of the crystal and m is an integer). In principle, an infinite plane-wave basis set is required to expand the electronic wavefunctions. However, the coefficients

$c_{i,\mathbf{k}+\mathbf{G}}$ for the plane waves with small kinetic energy $|\mathbf{k}+\mathbf{G}|^2/2$ are practically more important than those with large kinetic energy. Thus, the plane-wave basis set can be truncated to include only plane waves whose kinetic energies are less than some particular cutoff energy. The truncation of the plane-wave basis set at the finite cutoff energy will lead to an error in the computed total energy. However, it is possible to reduce the magnitude of the error by increasing the value of the cutoff energy.

When the plane waves are used as a basis set for the electronic wavefunctions, the Kohn-Sham equation is transformed into a simple form. Substitution of (2.30) into (2.4) and integration over r gives the secular equation written as

$$\sum_{\mathbf{G}'}\left[\frac{|\mathbf{k}+\mathbf{G}|^2}{2}\delta_{\mathbf{G}\mathbf{G}'}+v(\mathbf{G}-\mathbf{G}')+V_H(\mathbf{G}-\mathbf{G}')+V_{xc}(\mathbf{G}-\mathbf{G}')\right]c_{i,\mathbf{k}+\mathbf{G}'}=\varepsilon_i c_{i,\mathbf{k}+\mathbf{G}'},$$

(2.31)

where $v(\mathbf{G})$, $V_H(\mathbf{G})$, and $V_{xc}(\mathbf{G})$ are the Fourier transformed external, Hartree, and exchange-correlation potentials, respectively. In this form, the kinetic energy is diagonal, and the various potentials are described in terms of their Fourier transforms. The solution of (2.31) proceeds by diagonalization of the Hamiltonian matrix $H_{\mathbf{k}+\mathbf{G},\mathbf{k}+\mathbf{G}'}$ given by the terms in brackets. The size of the matrix is determined by the choice of cutoff energy $|\mathbf{k}+\mathbf{G}|^2/2$, and will be intractably large for the systems that contain both valence and core electrons. This is a severe problem, but it can be overcome by the use of pseudopotentials as explained in Sect. 2.1.3.

The calculations using the plane-wave basis set are in principle applied for periodic crystalline solids. However, the calculations can be approximately performed for the nonperiodic system containing a point defect or a crystal surface if we use the unit cell of periodic system as follows: The unit cell for a point defect in solids is schematically illustrated in Fig. 2.2a. The unit cell contains a defect

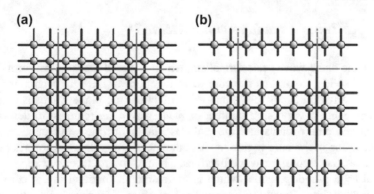

Fig. 2.2 Schematic illustration of calculation models for (**a**) point defect (vacancy) in a bulk solid, and (**b**) surface of a bulk solid. Circles and sticks represent atoms and bonds, respectively. Square represents the unit cell of calculation model

surrounded by a region of bulk crystal. Since the periodic boundary conditions are applied to the unit cell of periodic system, the energy per unit cell of the crystal containing an array of defects rather than that of a crystal containing an isolated defect is obtained. It is thus essential to incorporate enough bulk solid in the unit cell to eliminate the interaction of defects between the adjacent cells.

The unit cell for surfaces and interfaces is schematically illustrated in Fig. 2.2b. The unit cell contains a crystal slab and a vacuum region. Since the unit cell is repeated along the direction perpendicular to the crystal slab, the total energy of an array of crystal slab is calculated. To ensure that the results of the calculation accurately represent an isolated surface, the vacuum regions must be wide enough so that the adjacent crystal slabs do not interact through the vacuum region, and the crystal slabs must be thick enough so that the two surfaces of each crystal slab do not interact through the bulk crystal.

The Bloch's theorem changes the problem of calculating an infinite number of electronic wavefunctions to a finite number of electronic wavefunctions at an infinite number of **k** points. Since occupied states at each **k** point contribute to the electronic potential, an infinite number of calculations are necessary to compute this potential in principle. However, it is possible to represent the electronic wavefunctions in a region of **k** space by the wavefunction at a single **k** point in the region. In this case, the electronic states at only a finite number of **k** points are required to calculate the electronic potential, and hence determine the total energy. The calculations of the electronic states with special **k** points in the Brillouin zone [15, 16] enable us to obtain an accurate electronic potential and total energy of an insulator or a semiconductor with a very small number of **k** points. The electronic potential and the total energy of a metallic system are more difficult to obtain than those of insulator or semiconductor because a dense set of **k** points is required to define the fermi surface. However, the magnitude of error in the total energy due to inadequacy of the **k**-point sampling can be reduced by using a denser set of **k** points.

2.1.3 Ab Initio Pseudopotential Method

If we use the plane-wave basis formalism, which is one of the simplest formalism to implement solids, expanding the core wave functions or the core oscillatory region of valence wave functions into plane wave function into plane waves is extremely inefficient. Since a very large number of plane waves are required to describe the tightly bound core orbitals and the rapid oscillations of the wavefunctions of valence electrons in the core region, a vast amount of computational time is required to calculate the electronic wavefunctions. Hence, the replacement of an atomic potential by a pseudopotential which is more smooth function in the core region should reduce the magnitude of computational effort. The pseudopotential method allows the electronic wavefunctions to be expanded using a much smaller number of plane-wave basis set. Furthermore, it is well known that most of the physical properties of solids depend on valence electrons rather than core electrons. This pseudopotential exploits the characteristic features of valence electrons by removing the core electrons and replacing the strong ionic potential with a weaker pseudopotential that acts on a set of pseudo wavefunctions rather than the true valence wavefunctions, as shown in Fig. 2.3. The pseudopotential is constructed so that its scattering properties for the pseudo wavefunctions are identical to the scattering properties of the true potential for the real valence wavefunctions. Since a phase shift caused by the ion core is different for each angular momentum component of the valence wavefunction, the scattering from the pseudopotential must be angular momentum dependent. The most general form of a ionic pseudopotential V_{ion} is divided by a local part $V_{ion,local}^{PP}$ and a nonlocal part $V_{ion,nonlocal}^{PP}$

$$
\begin{aligned}
V_{ion} &= V_{ion,local}^{PP}(\boldsymbol{r}) + V_{ion,nonlocal}^{PP}(\boldsymbol{r}) \\
&= V_{ion,local}^{PP}(\boldsymbol{r}) + \sum_{lm} \big|lm\big\rangle V_{nonlocal,l}^{PP}(\boldsymbol{r}) \big\langle lm\big|,
\end{aligned}
\tag{2.32}
$$

where $|lm\rangle$ is the spherical harmonics and $V_{nonlocal,l}^{PP}(\boldsymbol{r})$ is the pseudopotential for angular momentum l. The scattering from the ion core is best described by a nonlocal pseudopotential that uses a different potential for each angular momentum component of the wavefunction. In the pseudopotential method, the Kohn-Sham equation using the plane-wave basis set shown in (2.31) is written as

Fig. 2.3 Schematic of model for solids with cores and valence electrons. The interacting atoms model can be evolved to a model of cores, which is composed of nucleus and core electrons, and valence electrons

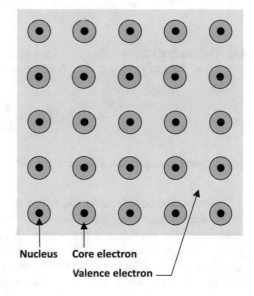

Nucleus   Core electron

Valence electron

$$\sum_{\mathbf{G}'} \left[\frac{|\mathbf{k} + \mathbf{G}|^2}{2} \delta_{\mathbf{GG}'} + V_{ion,local}^{PP}(\mathbf{G} - \mathbf{G}') + \sum_l V_{nonlocal,l}^{PP}(\mathbf{k} + \mathbf{G}, \mathbf{k} + \mathbf{G}') \right.$$
$$\left. + V_H(\mathbf{G} - \mathbf{G}') + V_{xc}(\mathbf{G} - \mathbf{G}') \right] c_{i,\mathbf{k}+\mathbf{G}'} = \varepsilon_i c_{i,\mathbf{k}+\mathbf{G}'}, \tag{2.33}$$

where the potential $V_{ion,local}^{PP}(\mathbf{G})$ is the Fourier transformed local potential and the potential $V_{nonlocal,l}^{PP}(\mathbf{G}, \mathbf{G}')$ is the momentum space representation of the nonlocal potential $|lm\rangle V_{nonlocal,l}^{PP}(r) \langle lm|$ [17, 18]. Various types of constructing schemes are introduced for nonlocal pseudopotentials, and they work extremely well. These nonlocal pseudopotentials are currently termed first-principles or norm-conserving.

The pseudopotentials currently used in the electronic structure calculation are generated from all-electron atomic calculation without any empirical parameters. The constructing procedure of these pseudopotential is to follow four general condition.

(1) The valence pseudo wave function generated from the pseudopotential should contain no nodes. This stems from the fact that we would like to construct smooth pseudo wave functions and therefore the undulations associated with nodes are undesirable.
(2) The normalized atomic pseudo wave function $R_l^{PS}(r)$ with angular momentum l and the real (all-electron) wave function $R_l^{AE}(r)$ agree beyond a chosen cutoff radius r_c which divides core and valence regions,

$$R_l^{PS}(r) = R_l^{AE}(r) \quad (r \geq r_c). \tag{2.34}$$

(3) The integrals from zero to r of the real and pseudo charge densities agree for $r \geq r_c$ for each valence state (norm conservation),

$$\int_0^{r_c} dr \left| R_l^{PS}(r) \right|^2 r^2 = \int_0^{r_c} dr \left| R_l^{AE}(r) \right|^2 r^2. \tag{2.35}$$

(4) The real and pseudo valence eigenvalues must be equal with each other.

The constructed pseudopotentials which fulfill the above conditions are generally called norm-conserving pseudopotential. The property of the norm conservation guarantees that electrostatic potential produced outside r_c is identical for real and pseudo charge distribution, and the scattering properties for the pseudo wave functions are identical to the scattering properties of the ion and the core electrons for the valence wave functions. There are many constructing schemes of norm-conserving pseudopotential. The pseudopotentials constructed by Troullier and Martins [19] are widely used in current pseudopotential calculations.

The Troullier-Martins pseudopotential is constructed by generalizing the Kerker procedure [20]. The defined radial part of the pseudo-wave-function $R_l^{PS}(r)$ is written as

$$R_l^{PS}(r) = \begin{cases} R_l^{PS}(r) & (r \geq r_c) \\ r^l \exp\{p(r)\} & (r < r_c), \end{cases} \tag{2.36}$$

where $p(r)$ is a six-order polynomial in r^2 expressed as

$$p(r) = \sum_{i=0}^{6} c_{2i} r^{2i}. \tag{2.37}$$

By inverting the radial Schrödinger equation, we can explicitly obtain the screened pseudopotential $V_l(r)$ as

$$V_l(r) = \begin{cases} V_{AE}(r) & (r \geq r_c) \\ \varepsilon_l + \frac{l+1}{2}\frac{p'(r)}{2} + \frac{p''(r) + \{p'(r)\}^2}{2} & (r < r_c), \end{cases} \tag{2.38}$$

where $V_{AE}(r)$ is the all electron potential, and ε_l is real eigenvalue with angular momentum l. The seven coefficients of the polynomial are determined from the following three conditions:

(1) Norm-conserving of charge within the core radius r_c,

$$2c_0 + \ln\left[\int_0^{r_c} dr r^{2(l+1)} \exp\{2p(r) - 2c_0\}\right] = \ln\left[\int_0^{r_c} dr |R_l^{AE}(r)|^2 r^2\right]. \tag{2.39}$$

(2) The continuity of the pseudo wave function, the pseudopotential, and their first two derivatives at r_c,

$$p(r_c) = \ln\left[\frac{r_c R_l^{AE}(r_c)}{r_c^{l+1}}\right], \tag{2.40}$$

$$p'(r_c) = \frac{r_c R_l^{AE}(r_c) + r_c R_l'^{AE}(r_c)}{r_c R_l^{AE}(r_c)} - \frac{l+1}{r_c}, \tag{2.41}$$

$$p''(r_c) = 2R_l^{AE}(r_c) - 2\varepsilon_l - \frac{2(l+1)}{r_c}p'(r_c) - [p'(r_c)]^2, \tag{2.42}$$

$$p(r_c)''' = 2R_l'^{AE}(r_c) + \frac{2(l+1)}{r_c^2}p'(r_c) - \frac{2(l+1)}{r_c}p''(r_c) - 2p'(r_c)p''(r_c), \quad (2.43)$$

$$p(r_c)'''' = 2R_l''^{AE}(r_c) - \frac{4(l+1)}{r_c^3}p'(r_c) + \frac{4(l+1)}{r_c^2}p''(r_c) - \frac{2(l+1)}{r_c}p'''(r_c)$$
$$- 2[p''(r_c)]^2 - 2p'(r_c)p'''(r_c), \quad (2.44)$$

where the primes denote differentiation with respect to r.

(3) The zero curvature of the screened pseudopotential at the origin, $V_l''(0) = 0$:

$$c_2^2 + c_4(2l+5) = 0. \quad (2.45)$$

The last condition gives smooth pseudopotential compared with the other norm-conserving pseudopotential such as Bachelet-Hamann-Schlüter type pseudopotential [21, 22]. The all-electron and pseudo wavefunctions and pseudopotentials obtained for silicon atoms are shown in Fig. 2.4.

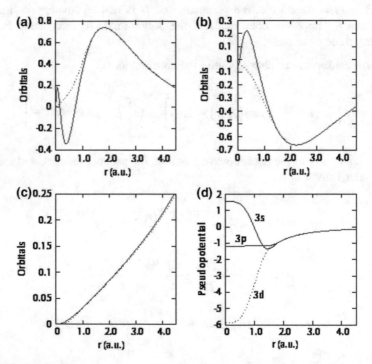

Fig. 2.4 Real (solid lines) and pseudo (dashed lines) radial wavefunction for **a** *3s*, **b** *3p*, and **c** *3d* orbitals and **d** angular components of pseudopotential for silicon as a function of radius obtained by the LDA. The cutoff radius r_c is 1.8 a.u

The local potential in (2.32) can be arbitrarily chosen. However, the summation in (2.32) will needed to be truncated at the same value of l, the local potential should be chosen so that it adequately reproduces the atomic scattering for all the higher angular momentum channels. The nonlocal potential $V_{ion,nonlocal}^{PP}(r)$ in (2.32) can be transformed into a separable form suggested by Kleinman and Bylander [23]

$$V_{nonlocal,l}^{KB}(r) = \frac{\left|V_{nonlocal,l}(r)\phi_l(r)\right\rangle\left\langle\phi_l(r)V_{nonlocal,l}(r)\right|}{\left\langle\phi_l(r)|V_{nonlocal,l}(r)|\phi_l(r)\right\rangle}, \tag{2.46}$$

where $\phi_l(r)$ is the atomic pseudo wavefunction with the angular momentum component. The final expression for the total energy within norm-conserving pseudopotential method is given by

$$E_{tot} = \sum_{\sigma i}\left\langle\phi_{\sigma i}\left|-\frac{1}{2}\nabla^2 + \sum_\tau V_{nlocal,l}^{KB}\right|\phi_{\sigma i}\right\rangle + \int dr\sum_\tau V_{ion,local}^{PP}(r - R_\tau)n(r)$$

$$+ \frac{1}{2}\int drdr'\frac{n(r)n(r')}{|r - r'|} + E_{xc}[n_\uparrow(r), n_\downarrow(r)], \tag{2.47}$$

where $|\phi_{l\sigma i}$ represents the i-th orbital with spin σ (= $\uparrow$, $\downarrow$), $n(r) = \sum_{\sigma i}\langle\phi_{\sigma i}|\phi_{\sigma i}\rangle$ is the total charge density, and τ is the index of atom situated at R_τ. The substantial advantage of this separable expression is saving the computer time and storage. When we use a plane wave basis set to express the wave functions (N_{PW} is the number of plane wave), the original pseudopotential in reciprocal space for angular momentum number l is written as $\sum_{i,j}|k + G_j\rangle V_{nonlocal,l}(k + G_j, k + G_i)\langle k + G_i|$. Here, $V_{nonlocal,l}(k + G_j, k + G_i)$ is expressed as

$$V_{nonlocal,l}(k + G_j, k + G_i) = \frac{2l+1}{4\pi\Omega}P_l(\cos\gamma)\int_0^\infty drV_{nonlocal,l}(r)j_l(|k + G_j|r)j_l(|k + G_i|r)r^2, \tag{2.48}$$

where $j_l(r)$ are spherical Bessel functions, $P_l(\cos\gamma)$ is the Legendre polynomials with

$$\cos\gamma - \frac{(k + G_i)(k + G_j)}{|k + G_i||k + G_j|}, \tag{2.49}$$

and Ω is the cell volume.

If the separable pseudopotential is used, $V_{nonlocal,l}(k + G_j, k + G_i)$ is transformed into $V_{nonlocal,l}^{KB}(k + G_j, k + G_i)$ expressed as,

$$
\begin{aligned}
V_{nonlocal,l}\big(\mathbf{k}+\mathbf{G}_j,\mathbf{k}+\mathbf{G}_i\big) = {} & \frac{1}{\Omega\langle\phi_l(r)|V_{nonlocal,l}(r)|\phi_l(r)\rangle} \\
& \times \left[\int_0^\infty dr\phi_l(r)V_{nonlocal,l}(r)j_l\big(|\mathbf{k}+\mathbf{G}_j|r\big)r^2\right] \\
& \times \left[\int_0^\infty dr\phi_l(r)V_{nonlocal,l}(r)j_l(|\mathbf{k}+\mathbf{G}_i|r)r^2\right] \\
& \times \sum_{m=-l}^{m} Y_{lm}\big(\mathbf{k}+\mathbf{G}_j\big)Y_{lm}^*(\mathbf{k}+\mathbf{G}_i),
\end{aligned}
\tag{2.50}
$$

where $Y_{lm}(\mathbf{G})$ are their spherical harmonics. As seen in (2.48) and (2.50), the number of plane wave integral for each l in the separable pseudopotential is N_{PW}, while that in the original pseudopotential is $N_{PW}(N_{PW}+1)/2$.

When there are tightly bound orbitals that have a substantial fraction of their weight inside the core region of the atom, the generation of softer (i.e., computationally faster) pseudopotentials within the framework of norm-conserving pseudopotentials is substantially difficult, and a high cutoff energy and large number of plane wave basis set is required to obtain accurate total energy. In order to generate much softer pseudopotentials, a more radical approach have been suggested by Vanderbilt, [24] in which the norm conservation condition in (2.35) is relaxed. These ultrasoft poseudopotentials are allowed to be as soft as possible, and the number of plane waves can be reduced dramatically. The generation algorithm of ultrasoft pseudopotentials also guarantees good scattering properties over a pre-specified energy range. This results in much better transferability and accuracy of the pseudopotentials.

2.2   Empirical Interatomic Potentials

Along with the development of the ab initio-based approach, empirical interatomic potentials for semiconductors have been proposed by many authors for application to dynamic treatments such as molecular dynamic treatment and for use in the simulation of complex systems with a large number of atoms, where the application of first-principles methods is currently limited by the large computational effort. Although the two-body interatomic potentials for ionic systems and metals is fairly well established, the theory of covalent solids is less developed because of the complexity of stabilizing the open tetrahedral structure of semiconductors, which require three-body potentials instead of simple two-body pairwise potentials such as Lennard-Jones and Morse potentials. The oldest empirical three-body potential for diamond or zinc blende structured semiconductors is the valence-force potential such as Keating model [25, 26], which has been used with considerable success for studying phonons, elastic properties, and excess energies of semiconductor alloy systems. However, it is perturbative in nature, and cannot properly be applied to systems with large distortion such as various crystal structures, defect structures, and surfaces. To treat these more general structures, Stillinger and Weber [27] successfully developed an empirical three-body interatomic potential to model melting of silicon, which consists of separable two- and three-body interactions. Another important approach closer to bond order form outlined by Abell [28], who noted that cohesive energies can be modeled by pairwise interactions moderated by the local environment in attempting to explain the universal cohesive energy curves for various materials, has been implemented by Tersoff [29] and Khor and Das Sarma [30].

Although these empirical interatomic potentials produced good global fits to cohesive energies for various crystal structures, elastic constants, and excess energies for semiconductors and their binary systems, very little systematic attempt has been made to obtain their empirical interatomic potentials for III-N semiconductors. Most of the potentials are developed for individual materials such as BN [31, 32] and GaN [33–36] using different formulations. This is because of the difficulty in universally reproducing the relative stability among threefold coordinated hexagonal (Hex, favored by BN), fourfold coordinated zinc blende (ZB, commonly appeared in III-V semiconductors), and wurtzite (WZ, stable in AlN, GaN, and InN) structures. Moreover, successful fabrications of various nanostructures such as nanowires, nanotubes, and nanocolumns require the development of empirical interatomic potentials applicable to systems with poorly coordinated atoms. However, the versatility of the potentials has not been carefully examined for poorly coordinated structures including Hex and a subtle energy difference between WZ and ZB crucial for III-N semiconductors. On the basis of these previous studies and the results obtained by the ab initio calculations, an empirical interatomic potential for III-N semiconductors is described within the framework of the bond order potential (BOP) [37, 38].

Using this BOP, the cohesive energy $D_e(r)$ of a given periodic structure is expressed as

$$D_e(r) = Z[Aq\exp(-\theta r) - Bp\exp(-\lambda r)], \tag{2.51}$$

where r is the interatomic distance, A, B, θ, and λ are constants, Z is the number of nearest neighbors, q is the number of valence electrons per atom ($q = 4$ for III-N semiconductors), and p is the bond order between two nearest atoms. A, B, θ, and λ are determined by reproducing the energy difference between ZB and rocksalt (RS) and the bulk modulus of the ZB phase. The bond order p is estimated as a function of Z analogous to the forms proposed by Bazant et al. [39] and Khor and Das Sarma [30] as follows:

$$p = \begin{cases} a\exp(-bZ^n) & (Z \leq 4) \\ (4/Z)^\alpha & (Z \geq 4), \end{cases} \tag{2.52}$$

where the parameters a, b, n, and α are determined to reproduce the energy differences between Hex and ZB, and between ZB and RS with conditions of $p = 1$ and the continuity of the first derivative of dp/dZ at $Z = 4$. The cohesive energies for the reference data of BN, AlN, GaN, and InN with Hex, ZB, and RS are

Table 2.3 Cohesive energy D_e and equilibrium interatomic distance r_e values for Hex, ZB, and RS in BN, AlN, GaN, and InN obtained by our ab initio calculations in [40]. Experimental results of r_e for WZ are also shown in parentheses

Structure		BN	AlN	GaN	InN
Hex	D_e (eV)	−6.885	−5.343	−4.261	−3.646
	r_e (Å)	1.446	1.807	1.864	2.088
ZB	D_e (eV)	−6.792	−5.832	−4.674	−4.180
	r_e (Å)	1.565	1.905	1.969	2.186
		(1.565)	(1.897)	(1.957)	(2.156)
RS	D_e (eV)	−5.010	−5.680	−4.265	−4.112
	r_e (Å)	1.749	2.031	2.122	2.336

Table 2.4 Potential parameter values of the Khor-Das Sarma and Bazant et al. functional form for BN, AlN, GaN, and InN determined in [38]

Parameter	BN	AlN	GaN	InN
A (eV)	410.550	497.513	714.627	743.464
B (eV)	579.681	313.590	924.370	1247.94
θ (Å^{-1})	3.87694	3.44767	3.25953	2.84488
λ (Å^{-1})	2.07703	1.71247	2.19928	2.14809
α	0.80386	0.53623	0.39877	0.25482
a	1.23219	1.18182	1.09035	1.04787
b	1.00×10^{-3}	1.95×10^{-3}	1.45×10^{-4}	2.45×10^{-5}
n	3.85	3.21	4.61	5.45
η	0.88830	0.66576	0.60996	0.52143
β	53.9478	47.0595	37.7707	29.9569
γ	3.15804	3.31636	3.27651	3.30710

obtained by ab initio total energy calculations [40]. Table 2.3 shows the cohesive energy (D_e) and equilibrium interatomic distance (r_e) values with experimental values of r_e at $Z = 4$ for BN [40–42], AlN, GaN, and InN [40, 43, 44]

Table 2.4 shows potential parameter values for BN, AlN, GaN, and InN in (2.51) and (2.52). By substituting values for a, b, and n in (2.52), the bond order $p^{(3)}$ at $Z = 3$ (corresponding to Hex) can be estimated to be 1.150 for BN, 1.106 for AlN, 1.066 for GaN, and 1.038 for InN. These $p^{(3)}$ values are favorably compared with those of C, BeO, Si, and AlP in the structural phase diagram of Hex with $Z = 3$ and WZ/ZB with $Z = 4$ in Fig. 2.5 [45]. The parameter $p^{(3)}$ and interatomic distance ratio $r_e^{(3)}/r_e^{(4)}$ in the phase diagram are employed owing to the fact that the cohesive energy ratio $D_e^{(3)}/D_e^{(4)}$ is simply described as a function of $p^{(3)} r_e^{(4)}/r_e^{(3)}$ to derive the phase boundary between $Z = 3$ and 4 using the criterion $D_e^{(3)}/D_e^{(4)} = 1$. This is based on the fact that the cohesive energy is proportional to a simple electrostatic interaction such as $D_e^{(Z)} \propto -Z p^{(Z)}/r_e^{(Z)}$ with $p^{(4)} = 1$ [45]. Here, BN is located in the stable region of Hex with $Z = 3$ similarly to C, while AlN, GaN, and InN remain in the stable region of WZ/ZB with $Z = 4$ similarly to BeO. It is found that Hex is

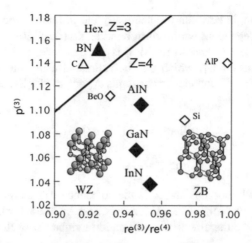

Fig. 2.5 Calculated $p^{(3)}$ and $r_e^{(3)}/r_e^{(4)}$ in BN, AlN, GaN, and InN denoted by closed triangle and closed diamonds in the phase diagram, respectively. The solid line indicates the phase boundary between threefold coordinated Hex and fourfold coordinated WZ/ZB. The data for C, BeO, Si, and AlP denoted by open triangle and open diamonds are also shown in this figure for comparison. Reproduced with permission from Ito et al. [38]. Copyright (2016) by the Japan Society of Applied Physics

favored by C and BN with a large $p^{(3)}$ at a small $r_e^{(3)}/r_e^{(4)}$, whereas WZ is more favorable in BeO, AlN, GaN, and InN with a small $r_e^{(3)}/r_e^{(4)}$ than ZB in Si and AlP with a large $r_e^{(3)}/r_e^{(4)}$. This is because the large $p^{(3)}$ and small $r_e^{(3)}$ give a strong attractive interaction between the bond charge located at the center of interatomic bonds and the positive charge located at the lattice sites that stabilizes poorly coordinated structures such as Hex. These results are consistent with experimental findings. Therefore, this empirical interatomic potential not only reproduces the energy difference but also gives physical insight into the relative stability between Hex and WZ/ZB.

To investigate the stability of various atomic arrangements including poorly coordinated structures, angular function should be incorporated in the form of empirical interatomic potential. This is because poorly coordinated atomic arrangements such as Hex, WZ, and ZB are unstable against shear distortion [26]. By employing the Khor-Das Sarma functional form in the angular term $G(\theta)$ [30], the empirical interatomic potential V_{ij} between i- and j-atoms is expressed as

$$V_{ij} = Aq\exp\left(-\theta r_{ij}\right) - Bp\exp\left(-\lambda r_{ij}\right)G(\theta), \qquad (2.53)$$

$$G(\theta) = 1 + \sum_{k(\neq i\,j)} \left[\cos\left(\eta\Delta\theta_{jik}\right) - 1\right], \qquad (2.54)$$

$$\Delta\theta_{jik} = \theta_{jik} - \theta_0(Z), \qquad (2.55)$$

where θ_{jik} is the angle between the nearest ij and jk interatomic bonds, and $\theta_0(Z)$ the coordination-dependent equilibrium bond angle (i.e., 120° at $Z = 3$, 109.47° at $Z = 4$, and 90° at $Z = 6$, and so forth). η is the parameter to be fitted to the bond bending force constant C_1, which can be estimated using the simple expression $C_1 = a^3(c_{11}\text{-}c_{12})/32$ [46] and elastic constants c_{11} and c_{12} for ZB estimated using the empirical rule [47] as

$$c_{11} = B_e\left(1 + \alpha_c^2\right), \tag{2.56}$$

$$c_{12} = B_e\left[1 - \left(\alpha_c^2/2\right)\right], \tag{2.57}$$

where B_e is the bulk modulus and α_c is the covalency. We estimate $C_1 = 5.808$ eV for BN, 2.680 eV for AlN, 2.673 eV for GaN, and 2.320 eV for InN. Khor and Das Sarma [30] also give the effective coordination number Z_i of the i-atom as

$$Z_i = \sum_{ij} \exp\left[-\beta(r_{ij} - R_0)^\gamma\right], \tag{2.58}$$

where R_0 is the minimum interatomic distance between i-atom neighbors. β and γ are determined by satisfying $Z_i = 5.9$ for β-tin and 9.7 for CsCl structures, respectively. The values of η, β, and γ are also listed in Table 2.4

Furthermore, an interaction beyond the second nearest neighbor (NN) is incorporated on the basis of the simple energy formula shown in (2.59) for the energy difference between WZ and ZB [48].

$$\Delta E_{WZ-ZB} = K\left[\frac{3}{2}(1 - f_i)\frac{Z_{bond}^2}{r_{bond}} - f_i\frac{Z_{ion}^2}{r_{ion}}\right], \tag{2.59}$$

where ΔE_{WZ-ZB} is the energy difference between WZ and ZB described as functions of the ionicity f_i, the covalent bond charge located at the center of the interatomic bond $Z_{bond}(= -2)$, the distance between covalent bond charges $r_{bond}(= c/2 \text{ in WZ})$, the ionic charge at the lattice site Z_{ion} (= 3 for III-N semiconductors), and the distance between ionic charges $r_{ion}(= 5c/8 \text{ in WZ})$. The value of 8.7 meV·Å is employed for K to reproduce $\Delta E_{WZ-ZB} = 25.3$ meV/atom for C with $f_i = 0$ obtained by ab initio calculations [49]. A simple criterion for WZ-ZB polytypism can be easily extracted as critical ionicity $f_i^c = 15/(15 + 2Z_{ion}^2)$ from $\Delta E_{WZ-ZB} = 0$ in (2.59). Here, WZ is a stable phase if $f_i \geq f_i^c$, whereas ZB is stable if $f_i < f_i^c$. The criterion implies that WZ is more stable than ZB when the ionicity is greater than 0.319 for group IV, 0.455 for III–V, 0.652 for II–VI semiconductors. Figure 2.6 shows the structural phase diagram for various compound semiconductors as functions of f_i and Z_{ion}, where the f_i values obtained by Pauling [50] are employed for 65 compounds [51]. This shows that the phase boundary between WZ and ZB is favorably compared with observations. Furthermore, this simple formulation has also been successfully applied to the polytypism in semiconductors such as SiC with 4H and 6H, in addition to 2H (WZ) and 3C (ZB) [52].

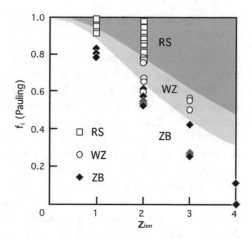

Thus, the interaction beyond the second NN corresponding to the electrostatic interaction V_{es} is simply given by

$$V_{es} = K \left[(1 - f_i) \frac{4}{r_{bond}} - f_i \frac{9}{r_{ion}} \right] f(r_{ion}). \qquad (2.60)$$

Here, we employ the ionicities of $f_i = 0.143$ for BN, 0.559 for AlN, 0.556 for GaN, and 0.649 for InN that give good estimates of ΔE_{WZ-ZB} consistent with ab initio calculations [48]. To apply V_{es} to any geometry, we choose the cutoff function $f(r_{ion})$ [39] to be exactly unity for WZ with gentle drop to zero for ZB as

$$f(r_{ion}) = \begin{cases} 1 & (r_{ion} < d) \\ \exp\left(\frac{\xi}{1 - x^{-3}}\right) & (d < r_{ion} < e) \\ 0 & (r_{ion} > e), \end{cases} \qquad (2.61)$$

where $d = 5c/8$, $e = \sqrt{33}c/8$, and $x = (r_{ion} - d)/(e - d)$. The parameter ξ is suitably determined by reproducing the properties such as energy change due to the rotation of the third NN atoms. The empirical interatomic potential V at $Z \sim 4$ is given by the summation of V_{ij} in (2.53) and V_{es} in (2.60). Table 2.5 shows ionicities and cohesive energies for III-N semiconductors with various crystal structures obtained by using (2.53) and (2.60).

Table 2.5 Ionicity f_i and cohesive energy D_e (eV) values for Hex, WZ, ZB, and RS in BN, AlN, GaN, and InN

	BN	AlN	GaN	InN
f_i	0.143	0.559	0.556	0.649
D_e [Hex]	−6.885	−5.343	−4.261	−3.646
D_e [WZ]	−6.775	−5.837	−4.680	−4.188
D_e [ZB]	−6.792	−5.832	−4.674	−4.180
D_e [RS]	−5.010	−5.680	−4.265	−4.112

2.3   Monte Carlo Simulations

In semiconductor technology, semiconducting materials are processed at high temperatures in the fabrication of thin films and devices. Thus, their fabrication involves the formation of thin films, alloys, defects, and related materials at high temperatures, since atomic motion or atomic exchange should be taken into account. The usual way to incorporate atomic motion and atomic exchange at realistic time duration is to use MC and MD methods. The MC method generally involves the use of random sampling techniques to estimate averages or to evaluate integrals. The MC method introduced by Metropolis et al. [53] is a very efficient and important sampling technique. For a constant density simulation, a system of N_{atom} particle is placed in an arbitrary initial configuration in a volume V, e.g., a lattice of chosen crystal packing and of uniform density equal to the experimental density at temperature T; here N_{atom}, V, and T are fixed. Configurations are generated according to the following rule: (i) Select particles at random; (ii) select random displacements or random exchanges of particles; (iii) calculate the change in potential energy ΔU after displacing or exchanging the chosen particle, where the ΔU is estimated by ab initio or empirical interatomic potential calculations; (iv) if ΔU is negative, accept the new configuration; (v) otherwise, select a random number h uniformly distributed over the interval $(0,1)$; (vi) if $\exp(-\Delta U/k_B T) < h$, accept the old configuration; (vii) otherwise, use the new configuration and the new potential energy as the current properties of the system. This procedure is repeated for a suitable number of configurations in order to approach equilibrium configurations. The quantities that can be evaluated by the Metropolis Monte Carlo (MMC) method are those that can be expressed as canonical averages of functions of configuration, such as pressure, energy, and the radial distribution function. The MMC method, however, does not generate a true dynamical history of an atomic system in contrast to the MD method which yields the atomic motion in the simulation duration to $t < 10^{-6}$ s because of its time-consuming numerical computations. Thus, the MMC method is suitable for achieving the equilibrium state but not for investigating non-equilibrium behavior such as adatom migration on the surface or thin-film formation.

A kinetic Monte Carlo (kMC) method based on the lattice-gas model is introduced to make up for the deficiencies in the MMC method without applicability to atomic motion and the MD methods with limited simulation duration. In the kMC simulation for atomic adsorption, diffusion, and desorption at specific lattice sites, individual atomic movements are simulated by the following procedure; (i) One first lists all the participating kinetic processes, and ascribes to each corresponding rate of occurrence. (ii) the kinetic rates are assumed to be of Arrhenius form such as $R = R_0 \exp(-\Delta E/k_B T)$, where R_0 is the prefactor in second and ΔE is the activation barrier of the kinetic events; (iii) once these rates have been calculated, the simulation proceeds to activate the kinetic events corresponding to these rates and to follow the individual atomic movements. The kMC method enables the simulation to finish within a realistic time $t \sim 1$ s for thin-film growth. In the kinetic events, each probability is exemplified by adsorption, diffusion, and desorption as follows. The site-correlated adsorption probability $P_{ad}(x)$ is written, assuming the local-thermal equilibrium approximation, by

$$P_{ad}(x) = \frac{\exp(-\Delta\mu(x)/k_B T)}{\{1 + \exp(-\Delta\mu(x)/k_B T)\}}. \tag{2.62}$$

Here $\Delta\mu(x)\left(= \mu_{ad}(x) - \mu_{gas}\right)$ is the difference between chemical potentials such as $\mu_{ad}(x)$ for an atom on the site x and μ_{gas} in the gas phase. The chemical potential of an atom on the surface $\mu_{ad}(x)$ corresponds to minus desorption energy $E_{ad}(x)$ and that in the gas phase μ_{gas} is given by

$$\mu_{gas} = -k_B T \ln\{g \times \varsigma_{trans} \times \varsigma_{rot} \times \varsigma_{vibr}\}, \tag{2.63}$$

where $\varsigma_{trans} = k_B T (2\pi m k_B T/h^2)^{3/2}/p_{gas}$, ς_{rot}, and ς_{vibr} are the partition functions for the translational motion, the rotational motion and the vibrational motion, respectively, g is the degree of degeneracy of the electron energy level (see Table 2.6), and p_{gas} is pressure [54, 55]. The diffusion probability $P_{diff}(x \rightarrow x')$ is assumed in the Arrhenius form of

$$P_{diff}(x \rightarrow x') = \nu_{lattice} \exp\left(-\frac{\Delta E(x \rightarrow x')}{k_B T}\right), \tag{2.64}$$

where diffusion prefactor $\nu_{lattice}$ is $2k_B T/h$ [56] and $\Delta E(x \rightarrow x')$ is the local activation energy involving the adatom hopping from site x to x'. The desorption probability $P_{de}(x)$ is written by

$$P_{de}(x) = \nu_{lattice} \exp\left\{-\left(\frac{E_{de}(x) - \Delta\mu(x)}{k_B T}\right)\right\}, \tag{2.65}$$

This equation implies that the difference $\Delta\mu(x)\left(= \mu_{ad}(x) - \mu_{gas}\right)$ affects the activation energy for desorption of the atom. That is, the adatom easily desorbs if μ_{gas} is lower than $\mu_{ad}(x)$, while the atom prefers to stay on the surface if μ_{gas} is higher than

Table 2.6 Electron energy level degeneracy g of some elements

Group	Element	g
I	H, Li, Na,K, Rb, Cs, Cu, Ag, Au	2
II	Be, Mg, Ca, Sr, Ba, Zn, Cd, Hg	1
III	B, Al, Ga, In, Tl	2
IV	C, Si, Ge, Sn, Pb	3
V	N, P, As, Sb, Bi	4
VI	O, S, Se, Te, Po	3
VII	F, Cl, Br, I	2
0	He, Ne, Ar, Kr, Xe, Rn	1

$\mu_{ad}(x)$. More precisely, the probability for surmounting the activation energy of $E_{de}(x)(= \exp(-\Delta E_{de}(x)/k_B T))$ is reduced (or enhanced) by a weighting function of $\exp(\Delta\mu(x)/k_B T)$ which corresponds to the local-thermal equilibrium desorption probability. On the basis of the above-mentioned stochastic differential equations, the kMC random-walk simulations are carried out to estimate lifetime τ and diffusion length L of one adatom on the surface. The diffusion coefficient D is also computed by

$$D = \frac{L^2}{2\tau}. \tag{2.66}$$

The applications of kMC method for initial growth processes on III-nitride semiconductor surfaces are described in Chap. 7.

References

1. P. Hohenberg, W. Kohn, Inhomogeneous electron gas. Phys. Rev. **136**, B864 (1964)
2. W. Kohn, L.J. Sham, Self-consistent equations including exchange and correlation effects. Phys. Rev. **140**, A1133 (1965)
3. J.P. Perdew, A. Zunger, Self-interaction correction to density-functional approximations for many-electron systems. Phys. Rev. B **23**, 5048 (1981)
4. D.M. Ceperley, B.J. Alder, Ground state of the electron gas by a stochastic method. Phys. Rev. Lett. **45**, 566 (1980)
5. J.P. Perdew, in *Electronic Structure of Solids '91*, ed. by P. Zeische, H. Eschrig (Academic, Berlin, 1991)
6. J.P. Perdew, K, Burke, M. Ernzerhof, Generalized gradient approximation made simple. Phys. Rev. Lett. **77**, 3365 (1996); **78**, 1396 (1997)
7. A.D. Becke, Density-functional exchange-energy approximation with correct asymptotic behavior. Phys. Rev. A **38**, 3098 (1988)
8. C. Lee, W. Yang, R.G. Parr, Development of the Colle-Salvetti correlation-energy formula into a functional of the electron density. Phys. Rev. B **37**, 785 (1988)
9. J. Heyd, G.E. Scuseria, M. Ernzerhof, Hybrid functionals based on a screened Coulomb potential. J. Chem. Phys. **118**, 8207 (2003)
10. M.T. Yin, M.L. Cohen, Theory of static structural properties, crystal stability, and phase transformations: application to Si and Ge. Phys. Rev. B **26**, 5668 (1982)
11. O.H. Nielsen, R.M. Martin, Stresses in semiconductors: *ab initio* calculations on Si, Ge, and GaAs. Phys. Rev. B **32**, 3792 (1985)
12. R.W. Godby, M. Schlüter, L.J. Sham, Accurate exchange-correlation potential for silicon and its discontinuity on addition of an electron. Phys. Rev. Lett. **56**, 2415 (1986)
13. M.S. Hybertsen, S.G. Louie, First-principles theory of quasiparticles: calculation of band gaps in semiconductors and insulators. Phys. Rev. Lett. **55**, 1418 (1985)
14. M.S. Hybertsen, S.G. Louie, Electron correlation in semiconductors and insulators: Band gaps and quasiparticle energies. Phys. Rev. B **34**, 5390 (1986)
15. A. Baldereschi, Mean-value point in the Brillouin zone. Phys. Rev. B **7**, 5212 (1973)
16. D.J. Chadi, M.L. Cohen, Special points in the Brillouin zone. Phys. Rev. B **8**, 5747 (1973)
17. J. Ihm, A. Zunger, M.L. Cohen, Momentum-space formalism for the total energy of solids. J. Phys. C: Solid State Phys. **12**, 4409 (1979)

18. M.C. Payne, M.P. Teter, D.C. Allen, T.A. Alrias, J.D. Joannopoulos, Iterative minimization techniques for *ab initio* total-energy calculations: molecular dynamics and conjugate gradients. Rev. Mod. Phys. **64**, 1045 (1992)
19. N. Troullier, J.L. Martins, Efficient pseudopotentials for plane-wave calculations. Phys. Rev. B **43**, 1993 (1991)
20. C.P. Kerker, Non-singular atomic pseudopotentials for solid state applications. J. Phys. **C13**, L189 (1980)
21. D.R. Hamman, M. Schlüter, C. Chiang, Norm-conserving pseudopotentials. Phys. Rev. Lett. **43**, 1494 (1979)
22. G.B. Bachelet, D.R. Hamman, M. Schlüter, Pseudopotentials that work: From H to Pu. Phys. Rev. B **26**, 4199 (1982)
23. L. Kleinman, D.M. Bylander, Efficacious form for model pseudopotentials. Phys. Rev. Lett. **48**, 1425 (1982)
24. D. Vanderbilt, Soft self-consistent pseudopotentials in a generalized eigenvalue formalism. Phys. Rev. B **41**, 7892 (1990)
25. P.N. Keating, Effect of invariance requirements on the elastic strain energy of crystals with application to the diamond structure. Phys. Rev. **145**, 637 (1966)
26. R.M. Martin, Elastic properties of ZnS structure semiconductors. Phys. Rev. B **1**, 4005 (1970)
27. F.H. Stillinger, T.A. Weber, Computer simulation of local order in condensed phases of silicon. Phys. Rev. B **31**, 5262 (1985)
28. G.C. Abell, Empirical chemical pseudopotential theory of molecular and metallic bonding. Phys. Rev. B **31**, 6184 (1984)
29. J. Tersoff, New empirical approach for the structure and energy of covalent systems. Phys. Rev. B **37**, 6991 (1988)
30. K.E. Khor, S. Das, Sarma, "Proposed universal interatomic potential for elemental tetrahedrally bonded semiconductors". Phys. Rev. B **38**, 3318 (1988)
31. W.H. Moon, M.S. Son, H.J. Hwang, Molecular-dynamics simulation of structural properties of cubic boron nitride. Phys. Rev. B **336**, 329 (2003)
32. W.H. Moon, H.J. Hwang, A modified Stillinger-Weber empirical potential for boron nitride. Appl. Surf. Sci. **239**, 376 (2005)
33. S.Q. Wang, Y.M. Wang, H.Q. Ye, A theoretical study on various models for the domain boundaries in epitaxial GaN films. Appl. Phys. A **70**, 475 (2000)
34. N. Aichoune, V. Potin, P. Ruteran, A. Hairie, G. Nouet, E. Paumier, An empirical potential for the calculation of the atomic structure of extended defects in wurtzite GaN. Comp. Mat. Sci. **17**, 380 (2000)
35. A. Béré, A. Serra, Atomic structure of dislocation cores in GaN. Phys. Rev. B **65**, 205323 (2002)
36. J. Nord, K. Albe, P. Erhart, K. Nordlund, Modelling of compound semiconductors: analytical bond-order potential for gallium, nitrogen and gallium nitride. J. Phys.: Condens. Matter **15**, 5649 (2003)
37. T. Ito, Recent progress in computer aided materials design for compound semiconductors. J. Appl. Phys. **77**, 4845 (1995)
38. T. Ito, T. Akiyama, and K. Nakamura, Systematic approach to developing empirical interatomic potentials for III–N semiconductors. Jpn. J. Appl. Phys. **55**, 05FM02 (2016)
39. M.Z. Bazant, E. Kaxiras, J.F. Justo, Environment-dependent interatomic potential for bulk silicon. Phys. Rev. B **56**, 8542 (1997)
40. Y. Takemoto, T. Akiyama, K. Nakamura, T. Ito, Theoretical study for crystal structure deformation in $A^N B^{8-N}$ compounds. e-J. Surf. Sci. Nanotechnol. **12**, 79 (2014)
41. M.B. Kaunoun, A.E. Merad, G. Merad, J. Cibert, H. Aourag, Prediction study of elastic properties under pressure effect for zincblende BN, AlN, GaN and InN. Solid-State Electron. **48**, 1601 (2004)
42. M. Grimsditch, E.S. Zouboulis, A. Polian, Elastic constants of boron nitride. J. Appl. Phys. **76**, 832 (1994)

43. A. Trampert, O. Brandt, K.H. Ploog, Crystal structure of group III Nitrides, in *Semiconductors and Semimetals*, ed. by J.I. Pankove, T.D. Mouskas, vol. 50, Chap. 7 (Academic Press, San Diego, 1998)
44. M.E. Sherwin, T.J. Drummond, Predicted elastic constants and critical layer thicknesses for cubic phase AlN, GaN, and InN on β-SiC. J. Appl. Phys. **69**, 8423 (1991)
45. T. Ito, T. Akiyama, K. Nakamura, Empirical interatomic potential approach to the stability of graphitic structure in $A^N B^{8-N}$ compounds. Jpn. J. Appl. Phys. **53**, 110304 (2014)
46. W.A. Harrison, *Electronic Structure and the Properties of Solids*, Chap. 8 (W. H. Freeman & Company, San Francisco, 1980)
47. S. Muramatsu, M. Kitamura, Simple expressions for elastic constants c_{11}, c_{12}, and c_{44} and internal displacements of semiconductors. J. Appl. Phys. **73**, 4270 (1993)
48. T. Ito, Simple criterion for wurtzite-zinc-blende polytypism in semiconductors. Jpn. J. Appl. Phys. **37**, L1217 (1998)
49. C.-Y. Yeh, Z.W. Lu, S. Froyen, A. Zunger, Zinc-blende–wurtzite polytypism in semiconductors. Phys. Rev. B **46**, 10086 (1992)
50. J.C. Phillips, Ionicity of the Chemical Bond in Crystals. Rev. Mod. Phys. **42**, 317 (1970)
51. T. Ito, T. Akiyama, K. Nakamura, Simple systematization of structural stability for $A^N B^{8-N}$ compounds. Jpn. J. Appl. Phys. **46**, 345 (2007)
52. T. Ito, T. Akiyama, K. Nakamura, A simple approach to the polytypism in SiC. J. Cryst. Growth **362**, 207 (2013)
53. N. Metropolis, A.W. Rosenbluth, M.N. Rosenbluth, A.H. Teller, E. Teller, Equation of state calculations by fast computing machines. J. Chem. Phys. **21**, 1087 (1953)
54. Y. Kangawa, T. Ito, A. Taguchi, K. Shiraishi, T. Ohachi, A new theoretical approach to adsorption–desorption behavior of Ga on GaAs surfaces. Surf. Sci. **493**, 178 (2001)
55. Y. Kangawa, T. Ito, Y.S. Hiraoka, A. Taguchi, K. Shiraishi, T. Ohachi, Theoretical approach to influence of As_2 pressure on GaAs growth kinetics. Surf. Sci. **507**, 285 (2002)
56. S. Clarke, D.D. Vvedensky, Origin of reflection high-energy electron-diffraction intensity oscillations during molecular-beam epitaxy: a computational modeling approach. Phys. Rev. Lett. **58**, 2235 (1987)

Chapter 3
Fundamental Properties of III-Nitride Compounds

Toru Akiyama

The reliability of calculated results using computational approach is crucial for discussing various aspects of growth related phenomena in III-nitride compounds. To ensure the calculated results, it is indispensable to check the validity of computational approach by comparing the experimental data. This is accomplished by comparing the calculated fundamental properties of III-nitride compounds, such as lattice parameters, cohesive energies, bulk modulus, band structures in bulk states, with those obtained by experiments. In this Chapter, we describe theses fundamental properties of III-nitride compounds obtained by computational approach. The calculated results of structural properties of BN, AlN, GaN, and InN by ab initio calculations are provided in Sect. 3.1. The calculated electronic properties of AlN, GaN, and InN with wurtzite structure, such as band structures, energy gap, and effective masses obtained by ab initio calculations are discussed in Sect. 3.2. In group III-nitride compounds, alloy semiconductor films such as InGaN and AlGaN are crucial to fabricate quantum wells (QWs) in light-emitting diodes (LEDs). We also discuss fundamental aspects of alloy semiconductors such as miscibility of InGaN obtained by empirical interatomic potential calculations in Sect. 3.3. Furthermore, dislocation core structures and its effects on the compositional inhomogeneity of InGaN and AlGaN obtained by empirical interatomic potential calculations are described in Sects. 3.4 and 3.5.

T. Akiyama (✉)
Department of Physics Engineering, Mie University, Tsu, Japan
e-mail: akiyama@phen.mie-u.ac.jp

© Springer International Publishing AG, part of Springer Nature 2018
T. Matsuoka and Y. Kangawa (eds.), *Epitaxial Growth of III-Nitride Compounds*,
Springer Series in Materials Science 269,
https://doi.org/10.1007/978-3-319-76641-6_3

3.1 Crystal Structure and Structural Stability

Figure 3.1 shows the crystal structures related to III-nitride compounds. There are many polytypes for BN, such as hexagonal (Fig. 3.1a), cubic zinc blende (ZB) (Fig. 3.1b), and wurtzite (WZ) (Fig. 3.1c) structures. The most stable structure in BN is the hexagonal structure shown in Fig. 3.1a The other III-nitride compounds, AlN, GaN, and InN take both ZB and WZ structures shown in Fig. 3.1b, c, respectively. The rocksalt structure shown in Fig. 3.1d is usually formed in oxide materials, but this structure is not stable for III-nitrides compared with the other structures, such as ZB and WZ structures.

The relative stability among these crystal structures can be confirmed by the calculated cohesive energies. Figure 3.2 shows the calculated cohesive energy of III-nitride compounds as a function of its bond length for various crystal structures. The most stable crystal structure for BN obtained by the cohesive energy shown in Fig. 3.2 is hexagonal structure, while that for AlN, GaN, and InN is WZ structure. It should be noted that the cohesive energy of ZB structure is close to that of WZ structure. For BN (Fig. 3.2a) the cohesive energy of ZB structure is 0.024 eV/atom lower than that of WZ structure, while those of WZ structure in AlN, GaN, and InN (shown in Fig. 3.2b–d, respectively) are lower than those of ZB structure by ~ 0.018 eV/atom. The calculated results indicate that both WZ and ZB can be formed as metastable structures in addition to most stable hexagonal structure in BN, and both ZB and WZ structures can be formed in AlN, GaN, InN. These calculated results for stable crystal structures of III-nitrides are consistent with the experimental results [1–3].

Table 3.1 summarize the calculated lattice constants, cohesive energies, bulk modulus B, and its first derivative with respect to the pressure B' for BN with hexagonal structure, and AlN, GaN, and InN with wurtzite structure obtained by ab initio calculations, along with experimental values [1–10]. The lattice parameters a and c are well reproduced by the calculations within the error of $\sim 3\%$: The calculated values are larger than the experimental values. This trend is commonly recognized in ab initio calculations using the GGA, in which lattice constants of semiconductors obtained by the GGA tend to overestimate the experimental values

(a)   **(b)**   **(c)**   **(d)**

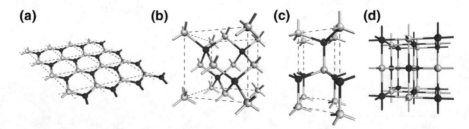

Fig. 3.1 Schematics of possible crystal structures in group III-V compound semiconductors **a** hexagonal, **b** zinc blende, **c** wurtzite, and **d** rocksalt structures. Filled and empty circles denote anion and cation, respectively. Dashed lines represents unit cell

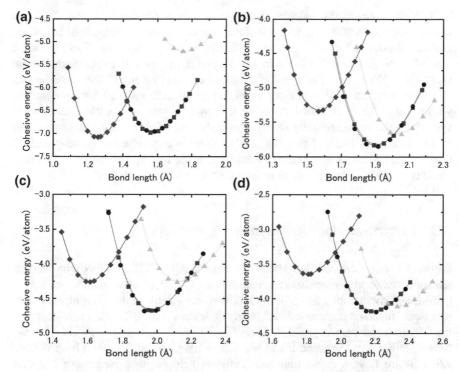

Fig. 3.2 Calculated cohesive energy as a function of bond length for various structures in **a** BN, **b** AlN, **c** GaN, and **d** InN. Squares, circles, diamonds, and triangles represent zinc blende, wurtzite, hexagonal, and rocksalt structures, respectively

Table 3.1 Calculated lattice constants a and c, cohesive energy E_{coh}, Bulk modulus B, and its first derivative with respect to the pressure B' for BN with hexagonal structure, and AlN, GaN, and InN with wurtzite structure obtained by ab initio calculations. Values in parentheses are experimental values

Material	a (Å)	c (Å)	E_{coh} (eV/atom)	B (10^{11} N/m^2)	B'
BN	2.51 (2.504[a])	6.78 (6.661[a])	−7.081 (−6.60[b])	2.27 (3.35[c])	3.5
AlN	3.12 (3.112[d])	5.08 (4.982[d])	−5.853 (−5.83[e])	1.90 (1.85[f])	3.6
GaN	3.22 (3.189[g])	5.25 (5.185[g])	−4.682 (−4.53[e])	1.71 (2.03[h])	4.2
InN	3.56 (3.451[i])	5.99 (5.760[i])	−4.195 (−3.99[e])	1.31 (1.65[j])	4.8

[a]Reference [1]
[b]Reference [2]
[c]Reference [3]
[d]Reference [4]
[e]Reference [5]
[f]Reference [6]
[g]Reference [7]
[h]Reference [8]
[i]Reference [9]
[j]Reference [10]

[11]. Consequently, the calculated elastic constants obtained by the GGA are usually underestimated. It is found that the calculated bulk modulus B of BN, GaN, and InN shown in Table 3.1 is smaller than those by experiments, [3, 8, 10] while that of AlN $(1.90 \times 10^{11} \text{ N/m}^2)$ is almost the same as the experimental value $(1.85 \times 10^{11} \text{ N/m}^2)$ [6]. The small difference between calculated and experimental values in wurtzite AlN originates from the small difference in lattice constants within 1.9%. The calculated cohesive energies of group-III nitrides reasonably agree with the experimentally reported cohesive energies, although the magnitude of cohesive energies are slightly larger than those in experiments. These calculated results suggest that ab initio calculations are feasible for investigating the structural stability of various semiconductors including III-nitride compounds.

3.2 Electronic Band Structure

Figure 3.3 shows the calculated band structure of AlN, GaN, and InN with wurtzite structure obtained by ab initio calculations. Both the valence band maximum (VBM) and conduction band minimum (CBM) are located at the Γ point, indicating that these materials possess a direct bandgap at the Γ point. This is consistent with the experimental results [12–14]. It should be noted that AlN with ZB structure have an indirect gap between Γ and X points while GaN and InN with both WZ and ZB structure have a direct transition between valence band maximum and conduction band minimum [11]. Furthermore, it is found that the calculated bandgap increases as group-III element becomes heavy. This trend can be intuitively understood by considering the energy levels of s orbitals ($3s$, $4s$, and $5s$ orbitals of Al, Ga, In atoms, respectively) of group-III elements, which mainly constitute the conduction bands. Table 3.2 shows the calculated gap energy of AlN, GaN, and InN with WZ structure from Fig. 3.3. The calculated gap energy can qualitatively reproduce the chemical trends of gap energy depending on the constituent group-III elements, but the calculated values are smaller than those obtained by experiments. It is well known that the GGA underestimate the bandgap of semiconductors, as explained in Sect. 2.1.

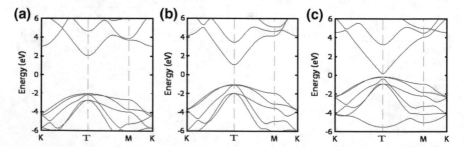

Fig. 3.3 Calculated band structure of **a** AlN, **b** GaN, and **c** InN with wurtzite structure. The Fermi energy is set at zero

Another important feature shown in Fig. 3.3 is the fine structure of valence bands at Γ point, since the band structures of WZ structure around the VBM are different from those of ZB structure. Figure 3.4 shows the schematics of fine structure of valence bands near Γ point in III-nitride compounds with WZ structure. The top of the valence band is split into twofold-degenerate and single states. The energy splitting between these states, which is labeled as crystal field splitting Δ_{cf}, is induced by the hexagonal symmetry of WZ structure. The value of Δ_{cf} depends on the kind of group-III elements. For AlN, the single state is higher than the twofold-degenerate state, and therefore Δ_{cf} is negative. On the other hand, for GaN and InN the twofold-degenerate state is higher than the single state, so that Δ_{cf} is positive, as also shown in Table 3.2. The difference between AlN and GaN agree with those in previous studies, [19] and the value of Δ_{cf} for GaN is reasonably consistent with the experimental value [16].

Furthermore, Table 3.2 summarizes the calculated electron and hole effective masses for bulk AlN, GaN, and InN with wurtzite structure obtained by ab initio calculations, along with available experimental data. The calculated value of electron mass for GaN normal to the [0001] direction (0.20 m_0, where m_0 is the free electron mass) agrees well with the experimental value of 0.23 m_0 [17]. The calculated hole masses have considerable **k**-direction and group-III element dependence. The masses of heavy-holes are ranging from 1.43 to 3.19. The masses of light- and crystal-holes have strong **k**-direction dependence. According to the experimental result for GaN, [18] the heavy-hole mass is 2.2 m_0. The calculated value is consistent with the experimental results. Although the calculated bandgaps in the GGA are underestimated, the GGA well reproduces the effective masses of group-III-nitride compounds.

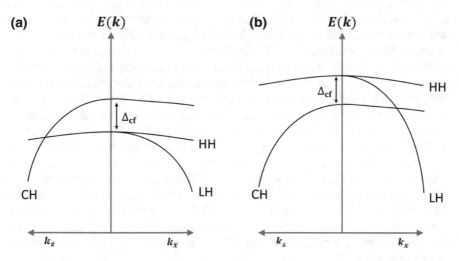

Fig. 3.4 Schematics of fine structure of valence bands near Γ point in group-III nitride compounds with wurtzite structure for **a** AlN and **b** GaN and InN. k_x (k_z) denotes reciprocal vector normal to the [1000] direction (along the [0001] direction). Values of crystal field splitting Δ_{cf} and effective masses of heavy-, light-, and crystal-hole bands (HH, LH, and CH, respectively) are listed in Table 3.2

Table 3.2 Calculated energy gap E_g (in eV), crystal field splitting Δ_{cf} (in meV), electron effective mass normal to the [0001] direction $m_{e\perp}$, electron effective mass along the [0001] direction $m_{e\parallel}$, hole effective mass normal to the [0001] direction $m_{h\perp}$, and hole effective mass along the [0001] direction $m_{h\parallel}$ for bulk AlN, GaN, and InN with wurtzite structure obtained by ab initio calculations. HH, LH, and CH denote the heavy-, light-, and crystal-hole bands, respectively. Values in parentheses are experimental data

| | E_g | Δ_{cf} | Electron effective mass | | Hole effective mass | | | | | |
| | | | | | HH | | LH | | CH | |
			$m_{e\perp}$	$m_{e\parallel}$	$m_{h\perp}$	$m_{h\parallel}$	$m_{h\perp}$	$m_{h\parallel}$	$m_{h\perp}$	$m_{h\parallel}$
AlN	4.07 (6.28[a])	−107 (−165[d])	0.32	0.29	3.19	1.97	0.31	1.97	3.10	0.23
GaN	2.11 (3.50[b])	44 (25[e])	0.20 (0.23[f])	0.17	2.50 (2.2[g])	2.27	0.19	2.27	1.03	0.15
InN	0.20 (0.78[c])	51	0.06	0.04	1.43	2.35	0.06	2.35	1.07	0.04

[a]Reference [12]
[b]Reference [13]
[c]Reference [14]
[d]Reference [15]
[e]Reference [16]
[f]Reference [17]
[g]Reference [18]

3.3 Miscibility of III-Nitride Alloy Semiconductors

Alloy semiconductor thin films exhibit various novel atomic arrangements such as atomic ordering and surface segregation [20–28]. Previous theoretical studies reveal that these novel atomic arrangements in thin films are closely related to the lattice constraint from the substrate. In III-nitride compounds, it has been reported that in InGaN grown GaN quantum dots (QDs) which enhance the lasing characteristic were spontaneously formed due to the compositional instability during their growth [29]. For developing optoelectronic devices such as laser diodes (LDs) using QDs, it is important to investigate the thermodynamic stabilities of the semiconductors. Saito and Arakawa [30] have investigated the phase stability of $In_xGa_{1-x}N$ based on a regular-solution model using an x-dependent interaction parameter Ω ($= -91.5x + 321$ meV) estimated from results of valence-force-field (VFF) calculation and obtained a slightly deviated miscibility gap from a symmetric one. However, these studies focused on the stabilities in bulk state and do not address those in the thin-film state. Since InGaN and GaN have a large lattice mismatch and therefore InGaN coherently grown on GaN is highly strained, the lattice constraint from the bottom layer must be incorporated to investigate thermodynamic stability in thin film state.

The thermodynamic stabilities of InGaN are assessed on the basis of excess energy calculations for InGaN thin films on GaN(0001) and InN(0001) substrates as well as those for the bulk InGaN. In the calculation procedure, empirical interatomic potentials described in Sect. 2.3 are used and lattice constraint from bottom layers such as GaN and InN is incorporated. The excess energy $\Delta E(x)$ is calculated by

$$\Delta E(x) = E(x) - \{xE(1.0) - (1-x)E(0.0)\}, \tag{3.1}$$

where $E(x)$, $E(1.0)$, and $E(0.0)$ are the system energies for $In_xGa_{1-x}N$, InN, and GaN, respectively.

Figure 3.5 shows the calculated excess energy as a function of composition obtained by using empirical interatomic potentials in Sect. 2.2 for bulk InGaN, InGaN/GaN, and InGaN/InN. The calculated excess energies have positive values over the entire composition range [31]. This indicates that a two-phase mixture is the most stable phase at 0 K. The excess energy curve for bulk InGaN shown in Fig. 3.5a is slightly deviated toward the Ga-rich side. For instance, the excess energy $\Delta E(0.25)$ is 63.6 meV/atom while $\Delta E(0.75)$ is 60.5 meV/atom. This is consistent with the VFF results [30]. As shown in Fig. 3.5b, c, we find that the excess energy maximum drastically shifts toward $x \sim 0.80$ for InGaN/GaN and $x \sim 0.10$ for InGaN/InN compared with $x \sim 0.50$ for bulk InGaN. Regarding InGaN/GaN, Karpov [32] have proposed that the critical temperature of phase separation shifted toward $x = 0.79$ compared with $x = 0.50$ for bulk, and the result using empirical interatomic potentials ($x \sim 0.80$) agrees with his result ($x = 0.79$). The origin of the shift is explained as follows. The bond length of GaN is smaller than that of InN. In case of InGaN/GaN, the epi-layer takes the planar compressive stress from the bottom layer, since the lattice constant of the epi-layer is larger than that of the bottom layer. Therefore, the In-rich epi-layer becomes less stable in contrast with the Ga-rich epi-layer. Consequently, the excess energy maximum shifts towards In-rich side. In the same manner, the excess energy maximum for InGaN/InN shifts toward the Ga-rich side due to the tensile stress accumulated in the epi-layer. These results indicate that the lattice constraint from the bottom layer influences the thermodynamic stabilities of thin films.

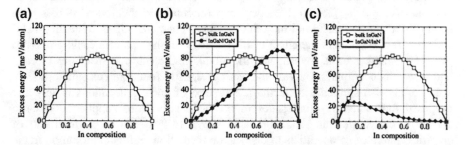

Fig. 3.5 Calculated excess energy as a function of indium composition: **a** bulk InGaN, **b** InGaN on GaN substrate (InGaN/GaN), and **c** InGaN on InN substrate (InGaN/InN). Reproduced with permission from Kangawa et al. [31]. Copyright (2000) by Elsevier

Furthermore, it should be noted that the excess energy for InGaN/GaN is larger than that for bulk at $x > 0.65$ shown in Fig. 3.5b. The result implies that the InGaN with large indium mole fraction is less stable on GaN layer than bulk state. This contra- dicts Karpov's result, [32] which shows that the InGaN/GaN is more stable than bulk state even in the range of $x > 0.65$. The discrepancy at large indium mole fraction seems to be caused by approximations employed in his calculations. In [32], the strain contribution to the thermodynamic stability in epitaxial state has been analytically estimated based on continuum theory, using lattice parameter a, elastic constant c_{ij} and interaction parameter W. In the calculations of continuum theory, the contribution of local atomic displacements in InGaN is neglected, while the calculations using interatomic potentials consistently incorporate the local atomic displacements. It has been clarified that the local atomic displacements strongly affect a, c_{ij} and W in III-V ternary alloys [33, 34]. Therefore, the discrepancy between the two results could be emphasized at large indium mole fraction, where the contribution of local atomic displacements becomes significant. On the other hand, the excess energy for InGaN/InN is smaller than that for bulk over almost the entire composition range, as shown in Fig. 3.5c. This implies that InGaN becomes more stable on InN layer than bulk state. These results suggest that the lattice constraint from the bottom layers has a significant influence on the thermodynamic stabilities of InGaN thin films grown on GaN and InN. It should be noted that the miscibility of AlGaN is different from that of InGaN because of the difference in lattice mismatch and cohesive energy. Since the difference between AlN and GaN is smaller than that between GaN and InN, it is expected that the excess energy of AlGaN is smaller that of InGaN. Indeed, the calculated excess energy obtained using interatomic potentials is at most 10 meV/atom.

3.4 Dislocation Core Structures

Since the lattice mismatch between the substrate such as sapphire and III-nitride compounds is quite large, a large number of threading dislocations ($\sim 10^8$–10^9 cm^{-2}) are inevitably generated during the epitaxial growth. These dislocations act as non-radiative enters, [35] so that many experimental and theoretical studies have been carried out to clarify not only the structural characteristic of threading dislocations [36–43] but also the effect of dislocations on the optoelectronic properties [43, 44]. Belabbas et al. investigated the atomic and electronic structures of edge and screw dislocations in GaN on the basis of density functional calculations [36]. Takei et al. also studied the electronic structure of edge dislocations in InN and proposed that dangling bond states in the dislocation core supply the electron carriers to the conduction band of bulk InN [37]. In this section, the feasibility of empirical interatomic potentials to dislocation formation and its contribution to the structural stability are discussed. The feasibility to dislocation formation is exemplified by the calculation of dislocation core energy, core radius and atomic coordinates for various dislocation core structures in wurtzite structured GaN and InN.

Figure 3.6 shows various dislocation core structures, such as those described by five- and seven-coordinated channels (5/7 core), four coordinated channels (4 core), and eight atom ring (8 core). The atomic structures of the dislocation core have been investigated by high-resolution transmission electron microscopy (HRTEM) [45] and by Z contrast [46]. Using empirical interatomic potentials described in Sect. 2.2, the system energy is calculated using a unit cell consisting of 7325 atoms for hypothetical wurtzite structured GaN and InN.

Using the system energy obtained by empirical interatomic potentials, the strain energy E_s stored in the cylinder of radius R is given by

$$E_s = \sum_{i=1}^{N(r_i < R)} \frac{E_{\text{dislocation}} - E_{\text{perfect}}}{r_i}, \tag{3.2}$$

where N is the number of atom within a cylinder containing the dislocation core, r_i the distance from the center of the cylinder to atom i, $E_{\text{dislocation}}$ and E_{perfect} are the system energies with and without dislocation, respectively. On the other hand, the linear elasticity theory [47] in an anisotropic crystal gives the strain energy as follows

$$E_s = A_0 \ln \left(\frac{R}{r_c} \right) + E_{\text{core}}, \tag{3.3}$$

where A_0 is the prelogarithmic factor, r_c the radius of dislocation core, R the radius of a cylinder containing the dislocation core, and E_{core} the dislocation core energy. Comparing the results obtained by (3.2) with (3.3), the dislocation core radius r_c can be estimated as the radius r_i where linear relationship between E_s and R is broken. The dislocation core energy E_{core} is determined by E_s at r_c.

Figure 3.7 shows the calculated strain energy in the cylinder of radius R as a function of $\ln R$ for AlN, GaN, and InN with various dislocation core structures [48, 49]. The region with linear relationship between energy and $\ln R$ corresponds to that satisfying linear elasticity theory [47]. On the other hand, the region deviating from

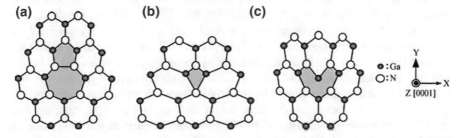

Fig. 3.6 Schematics of dislocation core structures of **a** 5- and 7-atom ring (5/7 core), **b** 4-atom ring (4 core), and **c** 8-atom ring (8 core) for wurtize III-nitride compounds along the [0001] direction. Shaded area denote dislocation core. Reproduced with permission from Kawamoto et al. [48]. Copyright (2005) by Elsevier

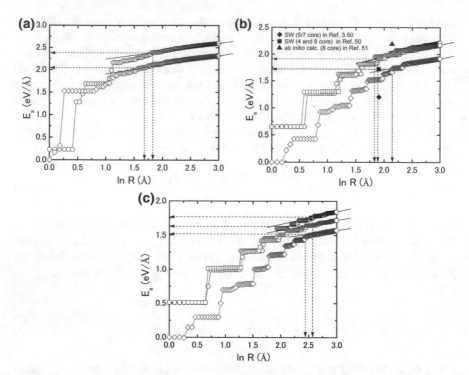

Fig. 3.7 Calculated energy stored in the cylinder of radius R as a function of $\ln R$ for wurtzite **a** AlN, **b** GaN and **c** InN. Open diamonds, squares, and circles denote the energies for 5/7, 4, and 8 core, respectively. Left- and down-arrows indicate core energy E_{core} and core radius r_c, respectively. Note that left-arrows for 4 core and 8 core in GaN and down-arrows for 5/7 core and 4 core in InN are degenerated. The dashed lines indicate the prelogarithmic factors. For GaN, data obtained by ab initio calculations and SW potentials are also shown

the linear relationship denotes the dislocation core region. The calculated results can be favorably compared with previously reported results obtained by the Stillinger-Weber (SW) potential and ab initio calculations [50, 51].

Table 3.3 lists the calculated core radius r_c, core energy E_{core} for AlN, GaN, and InN [48, 49]. The calculated results imply that the 5/7 core with the lowest E_{core} is the most stable. This is consistent with the results obtained by the SW potential for GaN. Furthermore, the prelogarithmic factor for the 5/7 core (A_0 = 0.48 eV/Å) for GaN is comparable to that in the linear elasticity theory (A_0 = 0.55 eV/Å). It is also find that the main distortion of the dislocation configurations in the present calculation is similar to that obtained by the SW potential. It should be noted that the E_{core} for InN (1.51 eV) is smaller than that of GaN (1.74 eV). This indicates that the 5/7 core structure consisting of InN is more favorable than that consisting of GaN. These calculated results thus suggest that the empirical interatomic potential calculations described in Sect. 2.2 are feasible for investigating the dislocation formation for wurtzite structured semiconductors such as GaN and InN as well as that for zinc blende structured semiconductors [52].

Table 3.3 Calculated core radius r_c and core energy E_{core} of edge dislocation with 5/7 core, 4 core, and eight core in GaN and InN. SW denotes the previously reported results for GaN obtained by Stillinger-Weber potential in[50]

Dislocation core type	r_c (Å)				E_{core} (eV/Å)			
	AlN	GaN	GaN (SW)	InN	AlN	GaN	GaN (SW)	InN
5/7 core	5.4	8.6	6.7	11.5	2.05	1.74	1.52	1.51
4 core		6.7	6.7	11.7		1.91	1.72	1.64
8 core	6.1	6.4	6.7	13.4	2.38	1.91	1. 72	1.77

3.5 Compositional Inhomogeneity Around Threading Dislocations

In the case of $In_{0.2}Ga_{0.8}N$ alloy semiconductors, the local concentration of In surrounding dislocations has been clarified by using empirical interatomic potentials and X-ray energy dispersive spectroscopy (EDX) [43]. In the experiments using TEM and EDX, Al segregation around threading dislocations in $Al_{0.3}Ga_{0.7}N$ have also been observed [41]. However, there have been few studies on the microscopic origin of the compositional inhomogeneity of group-III elements in AlGaN around threading dislocations from theoretical viewpoints. Furthermore, the difference in the compositional inhomogeneity of group-III elements around dislocations between AlGaN and InGaN alloys is unclear. In order to theoretically investigate the compositional inhomogeneity of group-III elements around threading dislocations, Monte Carlo (MC) simulations described in Sect. 2.3 using empirical interatomic potentials have been performed for the systems with dislocations in $Al_{0.3}Ga_{0.7}N$ and $In_{0.2}Ga_{0.8}N$ [53].

Figure 3.8 shows the schematic top views of calculation models used to simulate threading dislocations in $Al_{0.3}Ga_{0.7}N$ and $In_{0.2}Ga_{0.8}N$. We employ a unit cell with the size of $[50a \times 25\sqrt{3}a \times 2c]$ consisting of 20,000 atoms, where a and c are lattice constants along the [1000] and [0001] directions. Periodic boundary conditions are imposed along the [0001] direction, while fixed boundary conditions are applied in the [$1\bar{1}00$] and [$11\bar{2}0$] directions. For edge dislocation, the atomic configuration with the 5/7 core, which is the most stable core structure in GaN, has been considered [48, 49]. In the 5/7 core shown in Fig. 3.6a, the dislocation is generated by applying a displacement $\mathbf{u}^i = \left(u_x^i, u_y^i, 0\right)$ to each atom in the unit cell, expressed as [47]

$$u_x^i(x_i, y_i) = \frac{b_x}{2\pi} \left\{ \arctan\left(\frac{y_i}{r_i}\right) + \frac{\pi}{2}\text{sign}(y_i)[1 - \text{sign}(x_i)] + \frac{x_i y_i}{2(1-v)r_i^2} \right\}, \quad (3.4)$$

$$u_y^i(x_i, y_i) = \frac{b_x}{4\pi(1-v)} \left[-(1-2v)\ln r_i - \frac{x_i^2}{r_i^2} \right], \quad (3.5)$$

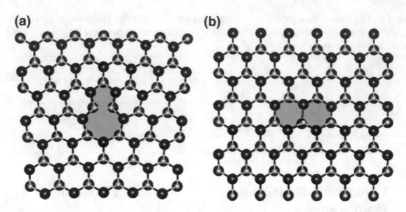

Fig. 3.8 Top views of calculation models of threading dislocations around **a** edge and **b** screw dislocation cores. Black and gray circles represent group III atoms and N atoms, respectively. Shaded area denotes the dislocation core. Reproduced with permission from Sakaguchi et al. [53]. Copyright (2016) by the Japan Society of Applied Physics

where b_x is the magnitude of the Burgers vector $\mathbf{b} = b_x \mathbf{e}_x$ in the $\left[11\bar{2}0\right]$ direction, ν is the Poisson ratio, x_i and y_i are the x and y components of the ith atom relative to the origin of the dislocation line location, and $r_i^2 = x_i^2 + y_i^2$. The Poisson ratios in $Al_{0.3}Ga_{0.7}N$ (0.270) and $In_{0.2}Ga_{0.8}N$ (0.288) are interpolated using the experimental values of AlN (0.266), GaN (0.271), and InN (0.354) [54–56]. For screw dislocations, the atomic configuration with a double 6-atom ring core is considered. This configuration was found to be the most stable in GaN in previous calculations [57]. In the double 6-atom ring core, the displacement is described by the vector $\mathbf{u}^i = \left(0, 0, u_z^i\right)$ with the component u_z^i given by [47]

$$u_z^i(x_i, y_i) = \frac{b_z}{2\pi}\left\{\arctan\left(\frac{y_i}{x_i}\right) + \frac{\pi}{2}\mathrm{sign}(y_i)[1 - \mathrm{sign}(x_i)]\right\}, \qquad (3.6)$$

where b_z is the magnitude of the Burgers vector $\mathbf{b} = b_z \mathbf{e}_z$ in the [0001] direction.

Figure 3.9 depicts the contour plots of Al composition around threading dislocations in $Al_{0.3}Ga_{0.7}N$. These contour plots indicate that the compositional inhomogeneity around threading dislocations depends on the type of dislocation. For edge dislocation, shown in Fig. 3.9a, Al atoms are preferentially located in one side of the dislocation core with compressive strain (upper region in Fig. 3.9a). On the other side of the dislocation core with tensile strain (lower region in Fig. 3.9a), Ga atoms are preferentially located. On the other hand, in the case of screw dislocation shown in Fig. 3.9b, Al atoms rarely appear in the lattice sites that form the dislocation core, and the composition of Al atoms excepting these lattice sites is almost constant at close to 0.3.

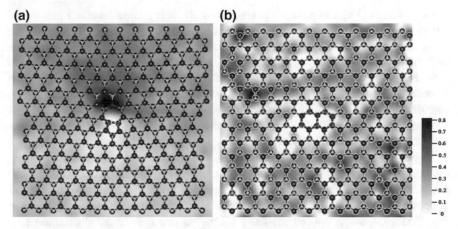

Fig. 3.9 Contour plots of Al composition around **a** edge and **b** screw dislocations in $Al_{0.3}Ga_{0.7}N$. The contour plots within 40×40 Å, which includes the dislocation core at the center, is shown. Black and gray circles represent group III elements and N atoms, respectively. Dashed lines denote the dislocation core. Reproduced with permission from Sakaguchi et al. [53]. Copyright (2016) by the Japan Society of Applied Physics

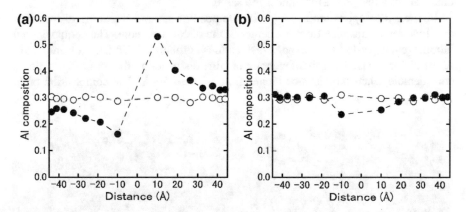

Fig. 3.10 Calculated Al composition as a function of distance from the center of **a** edge and **b** screw dislocations in $Al_{0.3}Ga_{0.7}N$. Positive and negative values of distance (horizontal axis) correspond to upper and lower regions in Fig. 3.9. Filled and empty circles represent values of equilibrium and initial (randomly distributed atomic arrangements) states, respectively. Reproduced with permission from Sakaguchi et al. [53]. Copyright (2016) by the Japan Society of Applied Physics

Figure 3.10 shows the averaged Al composition around threading dislocations as a function of distance from the center of the dislocation core. The compositional inhomogeneity of Al atoms in edge dislocations is quite different from that in screw dislocations. The inhomogeneity in edge dislocations shown in Fig. 3.10a is marked when the distance ranges from −30 to 30 Å. It should be noted that the Al composition at 10 Å is twice larger than that in the homogeneous case, whereas the Al composition at −10 Å is 60% smaller than that in the homogeneous case. In the case of screw dislocations, as shown in Fig. 3.10b, the Al composition is smaller than that in the inhomogeneous case when the distance is either −10 or 10 Å. This is because Al atoms rarely appear in the lattice site of the dislocation core. These calculated results suggest that the compositional inhomogeneity around edge dislocations is more marked than that around screw dislocations, which is qualitatively consistent with the EDX observation [41].

Figure 3.11 shows the contour plots of the In composition around threading dislocations in $In_{0.2}Ga_{0.8}N$. These contour plots indicate that, similar to $Al_{0.3}Ga_{0.7}N$, the compositional inhomogeneity around threading dislocations depends on the type of dislocation. In the case of the edge dislocations shown in Fig. 3.11a, Ga atoms are preferentially located in one side of the dislocation core with compressive strain (upper region in Fig. 3.11a), and In atoms are preferentially located on the other side of the dislocation core with tensile strain (lower region in Fig. 3.11a). On the other hand, as shown in Fig. 3.11b. In atoms rarely appear in the lattice sites that form the dislocation core, but are preferentially located near the dislocation core along the $[1\bar{1}00]$ direction. The composition of In atoms for the other lattice sites (~ 0.2) is almost the same.

Figure 3.12 depicts the averaged In composition around threading dislocations as a function of distance from the center of the dislocation core. The compositional inhomogeneity of In atoms in edge dislocations is found to be different from that in screw dislocations. The inhomogeneity in edge dislocations shown in Fig. 3.12a is considerable when the distance ranges from −30 to 30 Å. The composition of In

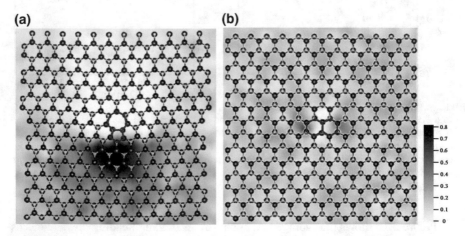

Fig. 3.11 Contour plots of In composition around **a** edge and **b** screw dislocations in $In_{0.2}Ga_{0.8}N$. The notations of circles and lines are the same as those in Fig. 3.9. Reproduced with permission from Sakaguchi et al. [53]. Copyright (2016) by the Japan Society of Applied Physics

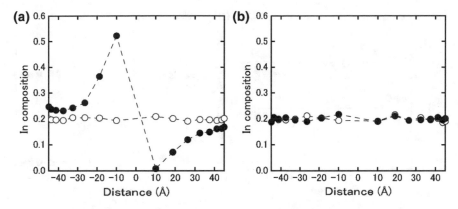

Fig. 3.12 Calculated In composition as a function of distance from the center of **a** edge and **b** screw dislocations in $In_{0.2}Ga_{0.8}N$. Positive and negative values of distance (horizontal axis) correspond to upper and lower regions in Fig. 3.11. The notations are the same as those in Fig. 3.10. Reproduced with permission from Sakaguchi et al. [53]. Copyright (2016) by the Japan Society of Applied Physics

atoms at -10 Å is $\sim 250\%$ larger than that in the homogeneous case, whereas the composition of In atoms at 10 Å is almost zero. The composition of In atoms in the case of screw dislocations shown in Fig. 3.12b does not show inhomogeneity of In atoms. However, as shown in Fig. 3.11b, a slight compositional inhomogeneity can be recognized along the $[1\bar{1}00]$ direction. Similar to the case of $Al_{0.3}Ga_{0.7}N$, the compositional inhomogeneity around edge dislocations in $In_{0.2}Ga_{0.8}N$ is more marked than that around screw dislocations.

Figure 3.13 shows the averaged bond lengths and interatomic potentials of Ga–N and Al–N bonds in edge dislocations of $Al_{0.3}Ga_{0.7}N$ in randomly distributed atomic arrangements represented by open circles in Fig. 3.10a and equilibrium states as a function of distance from the center of the dislocation core. Owing to the presence of both compressive—(upper region in Fig. 3.9a) and tensile—(lower region in Fig. 3.9b) strain regions around the dislocation core, bond lengths of Al–N and Ga–N in the region closer than -10 Å are quite different from those in the region farther than 10 Å. Since the Al–N bond length in the tensile-strain region (closer than -10 Å in Fig. 3.13a) is much larger than that in $Al_{0.3}Ga_{0.7}N$ without dislocations, the number of Al–N bonds in this region is reduced in the equilibrium state. This results in the lower interatomic potential values of Al–N bonds in the tensile-strain region closer than -10 Å shown in Fig. 3.13b. On the other hand, the number of Al–N bonds increases in the equilibrium state owing to shorter Al–N bonds in the compressive-strain region than that in $Al_{0.3}Ga_{0.7}N$ without dislocations. Consequently, the interatomic potentials of Al–N bonds decrease by 5.56 and 5.04 meV in the equilibrium state. The trend of bond distribution in $In_{0.2}Ga_{0.8}N$ is similar to that in $Al_{0.3}Ga_{0.7}N$. Since the bulk In–N bond length is markedly larger than

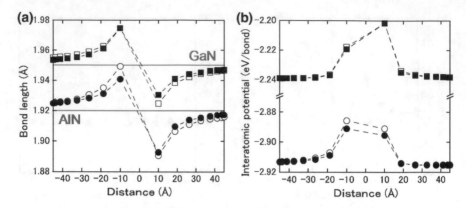

Fig. 3.13 Averaged **a** bond length and **b** interatomic potentials of Ga–N and Al–N bonds around edge dislocation in $Al_{0.3}Ga_{0.7}N$ as a function of distance from the center of dislocation core. Filled (empty) squares and filled (empty) circles represent the lengths of Ga–N and Al–N bonds in the equilibrium (initial) state, respectively. Horizontal lines represent averaged Ga–N (1.95 Å) and Al–N (1.92 Å) bond lengths in $Al_{0.3}Ga_{0.7}N$ without dislocation. Reproduced with permission from Sakaguchi et al. [53]. Copyright (2016) by the Japan Society of Applied Physics

the bulk Ga–N bond length, In atoms are placed in the tensile-strain region (lower region in Fig. 3.11a) so as to make the length of In–N bonds close to the In–N bond length in the bulk phase. In the screw dislocations, the distributions of Al–N and In–N bonds can also be understood from the stability of these bonds. Since bond lengths around the dislocation core are much longer than those in the bulk phase, Al–N and In–N bonds are stabilized by placing Ga and In atoms around screw dislocation cores in $Al_{0.3}Ga_{0.7}N$ and $In_{0.2}Ga_{0.8}N$, respectively.

References

1. R.S. Peace, An X-ray study of boron nitride. Acta Crystallogr. A **5**, 356 (1952)
2. P.K. Lam, R.M. Wentzcovitch, M.L. Cohen, High Density Phases of BN. Mater. Sci. Forum **54–55**, 165 (1990)
3. Y.N. Xu, W.Y. Ching, Electronic, optical, and structural properties of some wurtzite crystals. Phys. Rev. B **44**, 7787 (1991)
4. W. Yim, E. Stofko, P. Zanzucchi, J. Pankove, M. Ettenberg, S. Gilbert, Epitaxially grown AlN and its optical band gap. J. Appl. Phys. **44**, 292 (1973)
5. J. H. Edgar (ed.), *Properties of Group III Nitrides* EMIS Datareviews Series No. 11 (The Institution of Electrical Engineers, London 1994)
6. Q. Xia, H. Xia, A.L. Ruoff, Pressure-induced rocksalt phase of aluminum nitride: a metastable structure at ambient condition. J. Appl. Phys. **73**, 8198 (1993)
7. T. Detchprohm, K. Hiramatsu, K. Itoh, I. Akasaki, Relaxation process of the thermal strain in the GaN/α-Al₂O₃ heterostructure and determination of the intrinsic lattice constants of GaN free from the strain. Jpn. J. Appl. Phys. **31**, L1454 (1992)

8. D. Gerlich, S.L. Dole, G.A. Slack, Elastic properties of aluminum nitride. J. Phys. Chem. Solids **47**, 437 (1986)
9. T.L. Tansley, C.P. Foley, Optical band gap of indium nitride. J. Appl. Phys. **59**, 3241 (1986)
10. P.E. Van Camp, V.E. Van Doren, J.T. Devreese, Pressure dependence of the electronic properties of cubic III-V In compounds. Phys. Rev. B **41**, 1598 (1990)
11. C. Stampfl, C.G. Van de Walle, Density-functional calculations for III-V nitrides using the local-density approximation and the generalized gradient approximation. Phys. Rev. B **59**, 5521 (1999)
12. H. Yamashita, K. Fukui, S. Misawa, S. Yoshida, Optical properties of AlN epitaxial thin films in the vacuum ultraviolet region. J. Appl. Phys. **50**, 896 (1979)
13. B. Monemar, Fundamental energy gap of GaN from photoluminescence excitation spectra. Phys. Rev. B **10**, 676 (1974)
14. J. Wu, W. Walukiewicz, K.M. Yu, J.W. Ager III, E.E. Haller, H. Lu, W.J. Schaff, Y. Saito, Y. Nanishi, Unusual properties of the fundamental band gap of InN. Appl. Phys. Lett. **80**, 3967 (2002)
15. Y. Taniyasu, M. Kasu, Origin of exciton emissions from an AlN p-n junction light-emitting diode. Appl. Phys. Lett. **98**, 131910 (2011)
16. D.C. Reynolds, D.C. Look, W. Kim, Ö. Aktas, A. Botchkarev, A. Salvador, H. Morkoç, D.N. Talwar, Ground and excited state exciton spectra from GaN grown by molecular-beam epitaxy. J. Appl. Phys. **80**, 594 (1996)
17. M. Drechsler, D.M. Hofmann, B.K. Meyer, T. Detchprohm, H. Amano, I. Akasaki, Determination of the conduction band electron effective mass in hexagonal GaN. Jpn. J. Appl. Phys. **34**, L1178 (1995)
18. J.S. Im, A. Moritz, F. Steuber, V. Härle, F. Scholz, A. Hangleiter, Radiative carrier lifetime, momentum matrix element, and hole effective mass in GaN. Appl. Phys. Lett. **70**, 631 (1997)
19. A. Rubio, J.L. Corkill, M.L. Cohen, Quasiparticle band structure of AlN and GaN. Phys. Rev. B **48**, 11810 (1993)
20. K. Nakagawa, M. Miyao, Reverse temperature dependence of Ge surface segregation during Si-molecular beam epitaxy. J. Appl. Phys. **69**, 3058 (1991)
21. S. Fukatsu, K. Fujita, H. Yaguchi, Y. Shiraki, R. Ito, Self-limitation in the surface segregation of Ge atoms during Si molecular beam epitaxial growth. Appl. Phys. Lett. **59**, 2103 (1991)
22. N. Ohtani, S.M. Mokler, B.A. Joyce, Simulation studies of Ge surface segregation during gas source MBE growth of $Si/Si_{1-x}Ge_x$ heterostructures. Surf. Sci. **295**, 325 (1993)
23. D.E. Jesson, S.J. Pennycook, J.-M. Baribeau, D.C. Houghton, Atomistic processes of surface segregation during Si/Ge MBE growth. Thin Solid Films **222**, 98 (1992)
24. T.S. Kuan, T.F. Kuech, W.I. Wang, E.L. Wilkie, Long-range order in $Al_xGa_{1-x}As$. Phys. Rev. Lett. **54**, 201 (1985)
25. A. Gomyo, T. Suzuki, K. Kobayashi, S. Kawata, I. Hino, T. Yuasa, Evidence for the existence of an ordered state in $Ga_{0.5}In_{0.5}P$ grown by metalorganic vapor phase epitaxy and its relation to band-gap energy. Appl. Phys. Lett. **50**, 673 (1987)
26. O. Ueda, M. Takikawa, J. Komeno, I. Umebu, Atomic structure of ordered InGaP crystals grown on (001)GaAs substrates by metalorganic chemical vapor deposition. Jpn. J. Appl. Phys. **26**, L1824 (1987)
27. N. Grandjean, J. Massies, S. Dalmasso, P. Vennegues, L. Siozade, L. Hirsch, GaInN/GaN multiple-quantum-well light-emitting diodes grown by molecular beam epitaxy. Appl. Phys. Lett. **74**, 3616 (1999)
28. N. Duxbury, U. Bangert, P. Dawson, E.J. Thrush, W. Van der Stricht, K. Jacobs, I. Moerman, Indium segregation in InGaN quantum-well structures. Appl. Phys. Lett. **76**, 1600 (2000)
29. Y. Narukawa, Y. Kawakami, M. Funato, Sz. Fujita Sg. Fujita, S. Nakamura, Role of self-formed InGaN quantum dots for exciton localization in the purple laser diode emitting at 420 nm. Appl. Phys. Lett. **70**, 981 (1997)

30. T. Saito, Y. Arakawa, Atomic structure and phase stability of $In_xGa_{1-x}N$ random alloys calculated using a valence-force-field method. Phys. Rev. B **60**, 1701 (1999)
31. Y. Kangawa, T. Ito, A. Mori, A. Koukitu, Anomalous behavior of excess energy curves of $In_xGa_{1-x}N$ grown on GaN and InN. J. Cryst. Growth **220**, 401 (2000)
32. S.Y. Karpov, Suppression of phase separation in InGaN due to elastic strain. MRS Int. J. Nitride Semicond. Res **3**, 16 (1998)
33. T. Ito, A pseudopotential approach to the structural and thermodynamical properties of III–V ternary semiconductor alloys. Phys. Status Solidi B **129**, 559 (1985)
34. T. Ito, A pseudopotential approach to the disorder effects in III-V ternary semiconductor alloys. Phys. Status Solidi B **135**, 493 (1986)
35. F.S.D. Lester, F.A. Ponce, M.G. Craford, D.A. Steigerwald, High dislocation densities in high efficiency GaN-based light-emitting diodes. Appl. Phys. Lett. **66**, 1249 (1995)
36. I. Belabbas, J. Chen, G. Nouet, Electronic structure and metallization effects at threading dislocation cores in GaN. Comput. Mater. Sci. **90**, 71 (2014)
37. Y. Takei, T. Nakayama, Electron-carrier generation by edge dislocations in InN films: First-principles study. J. Cryst. Growth **311**, 2767 (2009)
38. F.A. Ponce, S. Srinivasan, A. Bell, L. Geng, R. Liu, M. Stevens, J. Cai, H. Omiya, H. Marui, S. Tanaka, Microstructure and electronic properties of InGaN alloys. Phys. Status Solidi B **240**, 273 (2003)
39. N. Duxbury, U. Bangert, P. Dawson, E. Thrush, W. Van der Stricht, K. Jacobs, I. Moerman, Indium segregation in InGaN quantum-well structures. Appl. Phys. Lett. **76**, 1600 (2000)
40. H. Lei, J. Chen, P. Ruterana, Influences of the biaxial strain and c-screw dislocation on the clustering in InGaN alloys. J. Appl. Phys. **108**, 103503 (2010)
41. L. Chang, S.K. Lai, F.R. Chen, J.J. Kai, Observations of Al segregation around dislocations in AlGaN. Appl. Phys. Lett. **79**, 928 (2001)
42. C.J. Fall, R. Jones, P.R. Briddon, A.T. Blumenau, T. Frauenheim, M.I. Heggie, Influence of dislocations on electron energy-loss spectra in gallium nitride. Phys. Rev. B **65**, 245304 (2002)
43. M.K. Horton, S. Rhode, S.L. Sahonta, M.J. Kappers, S.J. Haigh, T.J. Pennycook, C. J. Humphreys, R.O. Dusane, M.A. Moram, Segregation of Into Dislocations in InGaN. Nano Lett. **15**, 923 (2015)
44. Y. Mera, K. Maeda, Optoelectronic activities of dislocations in gallium nitride crystals. IEICE Trans. Electron. **E83-C**, 612 (2000)
45. V. Potin, P. Ruterana, G. Nouet, R.C. Pond, H. Morkoç, Mosaic growth of GaN on (0001) sapphire: A high-resolution electron microscopy and crystallographic study of threading dislocations from low-angle to high-angle grain boundaries. Phys. Rev. B **61**, 5587 (2000)
46. Y. Xin, S.J. Pennycook, N.D. Nellist, S. Sivanathan, F. Ommnes, B. Neaumont, J.P. Faurie, P. Gibart, Direct observation of the core structures of threading dislocations in GaN. Appl. Phys. Lett. **72**, 2680 (1998)
47. J.P. Hirth, J. Lothe, *Theory of dislocations* (Wiley, New York, 1982)
48. K. Kawamoto, T. Suda, T. Akiyama, K. Nakamura, T. Ito, An empirical potential approach to dislocation formation and structural stability in GaN_xAs_{1-x}. Appl. Surf. Sci. **244**, 182 (2005)
49. T. Ito, S. Inahama, T. Akiyama, K. Nakamura, Systematic theoretical investigations of compositional inhomogeneity in $In_xGa_{1-x}N$ thin films on GaN(0001). J. Cryst. Growth **298**, 186 (2007)
50. A. Bere, A. Serra, Atomic structure of dislocation cores in GaN. Phys. Rev. B **65**, 205323 (2002)
51. S.M. Lee, A. Belkhir, X.Y. Zhu, Y.H. Lee, Y.G. Hwang, Th Frauenheim, Electronic structures of GaN edge dislocations. Phys. Rev. B **61**, 16033 (2000)
52. N. Miyagishima, T. Shinoda, K. Suzuki, T. Kaneko, K. Takeda, K. Shiraishi, T. Ito, Atomic and electronic structure of misfit dislocations in GaSb/GaAs(001). Phys. B **340–342**, 1009 (2003)

53. R. Sakaguchi, T. Akiyama, K. Nakamura, T. Ito, Theoretical investigations of compositional inhomogeneity around threading dislocations in III–nitride semiconductor alloys. Jpn. Appl. Phys. **55**, 05FM05 (2016)
54. A. Polian, M. Grimsditch, I. Grzegory, Elastic constants of gallium nitride. J. Appl. Phys. **79**, 3343 (1996)
55. L.E. McNeil, M. Grimsditch, R.H. French, Vibrational Spectroscopy of Aluminum Nitride. J. Am. Ceram. Soc. **76**, 1132 (1993)
56. A.U. Sheleg, V.A. Savastenko, Determination of elastic constants of hexagonal crystals from measured values of dynamic atomic displacements. Izv. Akad. Nauk SSSR Neorg. Mater. **15**, 1598 (1979). [in Russian]
57. I. Belabbas, J. Chen, G. Nouet, A new atomistic model for the threading screw dislocation core in wurtzite GaN. Comput. Mater. Sci. **51**, 206 (2012)

Chapter 4
Fundamental Properties of III-Nitride Surfaces

The control of growth conditions is one of the important factors for fabricating high-quality crystals and would be achieved through the understanding of surface reconstructions. It is well known that reconstructed structures appear on the growth front (surfaces) of semiconductor materials, so that investigations for the reconstructions on III-nitride surfaces are necessary from theoretical viewpoints taking growth conditions into account. Surface energy calculations for various surface structures using ab initio calculations have revealed that the stable surface reconstruction depends on the chemical potential of constituent atomic species [1–6]. Although these ab initio studies successfully elucidate some aspects in the surface-related problems, their results are limited at 0 K without incorporating the growth parameters such as temperature and beam equivalent pressure (BEP) during molecular beam epitaxy (MBE) growth. In order to make up the deficiency in the previously reported ab initio calculations, an ab initio-based approach to include temperature and BEP has been proposed and successfully applied to the surface reconstructions and elemental growth processes on the GaAs and InAs surfaces [7–10]. In this Chapter, recent achievements for clarifying the reconstruction on III-nitride surfaces including polar, nonpolar and semipolar surfaces using the ab initio-based approach are described. Surface phase diagram calculations as functions of temperature and BEP are performed for surfaces exemplified by those with polar (0001) and (000$\bar{1}$), nonpolar (1$\bar{1}$00) and (11$\bar{2}$0), and semipolar (1$\bar{1}$01) and (11$\bar{2}$2) orientations, shown in Fig. 4.1 [10–16]. The role of hydrogen adsorption in surface phase diagrams is also investigated in conjunction with metal-organic vapor-phase epitaxy (MOVPE) growth [17–26].

T. Akiyama (✉)
Department of Physics Engineering, Mie University, Tsu, Japan
e-mail: akiyama@phen.mie-u.ac.jp

© Springer International Publishing AG, part of Springer Nature 2018
T. Matsuoka and Y. Kangawa (eds.), *Epitaxial Growth of III-Nitride Compounds*,
Springer Series in Materials Science 269,
https://doi.org/10.1007/978-3-319-76641-6_4

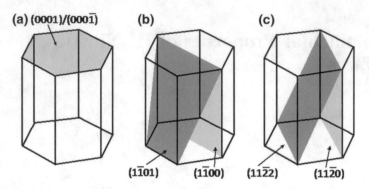

Fig. 4.1 Schematics of crystal planes such as **a** polar $(0001)/(000\bar{1})$, **b** nonpolar $(1\bar{1}00)$ and semipolar $(1\bar{1}01)$, **c** nonpolar $(11\bar{2}0)$ and semipolar $(11\bar{2}2)$ orientations in wurtzite structured AlN, GaN, and InN

4.1 Surface Phase Diagram Calculations

The relative stability among various surface structures including hydrogen (in the case of GaN) is determined using the formation energy E_f given by

$$E_f = E_{tot} - E_{ref} - \sum_i n_i \mu_i, \tag{4.1}$$

where E_{tot} and E_{ref} are the total energy of the surface under consideration and of the reference surface, respectively, μ_i is the chemical potential of the ith species, and n_i is the number of excess or deficit ith atoms with respect to the reference. Here, we assume that the surface is in equilibrium with bulk GaN expressed as

$$\mu_{Ga} + \mu_N = \mu_{GaN}, \tag{4.2}$$

where μ_{GaN} is the chemical potential of bulk GaN. μ_{Ga} can vary in the thermodynamically allowed range $\mu_{Ga}^{bulk} + \Delta H_f \leq \mu_{Ga} \leq \mu_{Ga}^{bulk}$, where ΔH_f is the heat of formation of bulk GaN (μ_{Ga}^{bulk} is the chemical potential of bulk Ga). The lower and upper limits correspond to N-rich and Ga-rich conditions, respectively. The same formalism is also applied to the study of AlN and InN surfaces using the chemical potentials of bulk Al (μ_{Al}^{bulk}) and bulk In (μ_{In}^{bulk}) as functions of Al and In chemical potentials, respectively. The calculated values of ΔH_f obtained by the generalized gradient approximation (GGA) are -2.91, -1.24, and -0.37 eV for AlN, GaN, and InN, respectively.

The reconstruction under growth conditions, such as temperature and pressure, is determined using the surface phase diagrams. The concept of surface phase diagram calculations is shown in Fig. 4.2. The surface phase diagrams are obtained by comparing the free energy or chemical potential of an atom/molecule in solid phase (μ_{solid}) with that of in gas phase (μ_{gas}). The free energy per particle (chemical

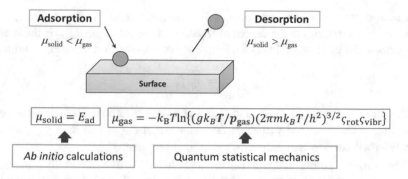

Fig. 4.2 Schematics of surface phase diagram calculation on the basis of ab initio calculations. By comparing the values of chemical potential, μ_{gas}, with adsorption energy, E_{ad}, obtained by ab initio calculations, the adsorption-desorption behavior of an adatom/molecule can be determined

potential) in gas phase is written in (2.63), which can be computed using quantum statistical mechanics. The free energy in solid phase corresponding to the adsorption energy ($E_{ad} = \mu_{solid}$) can be obtained using ab initio calculations. The adsorption energy considered is the energy difference between the two slab models. One model is a surface with an adatom, and the other is a surface without an adatom, i.e., the adatom is in the vacuum region. The adsorption-desorption activation energy is not considered because we consider the static behavior is considered to construct the surface phase diagrams. However, the activation energy should be considered if the growth kinetics are investigated as discussed in Chap. 7. By comparing μ_{gas} with E_{ad}, we can discuss the adsorption-desorption behavior, as shown in Fig. 4.2. The free energy of vibrational contribution is very small compared with the energy difference between a given structure and the ideal surface [27, 28]. Thus, when the temperature or pressure is varied, the gas-phase entropy difference is also considerably larger than the surface entropy change. Therefore, only the entropic effects of the gas phase are considered in surface phase diagram calculations. In the gas phase chemical potential, the partition functions in (2.63) are written as

$$\varsigma_{trans} = \frac{k_B T}{p_{gas}} \left(2\pi m k_B T / h^2\right)^{3/2}, \tag{4.3}$$

$$\varsigma_{rot} = \frac{1}{\pi \sigma} \left\{ \frac{8\pi^3 (I_1 I_2 \cdots)^{1/n_{rot}} k_B T}{h^2} \right\}^{n_{rot}/2}, \tag{4.4}$$

$$\varsigma_{vibr} = \prod_i^{3N_{atom}-3-n_{rot}} \frac{1}{1 - \exp\left(-h v_i / k_B T\right)}, \tag{4.5}$$

where m is the mass of one particle, σ is symmetric factor, I_j $(j = 1, 2, \ldots)$ is the moment of inertia, n_{rot} is the degree of freedom of the rotation, N_{atom} is the number of atoms in the particle, and v_i is the frequency of ith vibration mode. I_j is written as

$$I_j = m_j r_j^2, \tag{4.6}$$

where m_j is the reduced mass, and r_j is the gyration radius. The structure corresponding to the adsorbed surface is favorable when E_{ad} is less than μ_{gas}, whereas the desorbed surface is stabilized when μ_{gas} is less than E_{ad}.

The surface phase diagram calculations have been successfully applied to determine the reconstructions on GaAs(001) surfaces, [7] which are experimentally observed under Ga-rich conditions. Figure 4.3 shows the calculated gas phase chemical potential of Ga atom as a function of temperature for various Ga-BEP. It is found that the calculated gas phase chemical potential decreases with temperature and increases with BEP. By comparing the calculated adsorption energy of Ga atom ($E_{ad} = -3.3$ eV), The gas phase chemical potential becomes lower than E_{ad} when temperatures is higher than 1000 K for the condition of Ga-BEP at 1.0×10^{-5} Torr. This suggests that the critical temperature for Ga adsorption is ~ 1000 K for Ga BEP of 1.0×10^{-5} Torr. The critical temperatures for Ga adsorption under various BEP conditions are plotted in Fig. 4.4. The surface phase diagram for Ga adsorption thus suggest that the Ga-droplet appear under low temperature and high Ga-BEP conditions, while GaAs-(2×4) $\beta2$ surface is stabilized under high temperature and low Ga-BEP conditions. Furthermore, the calculated surface phase diagram agrees with the experimental findings, where Ga-droplets are observed under ~ 900 K during the MBE growth of GaAs under Ga-rich conditions [29] and Ga desorption proceeds above ~ 970 K after turning off the Ga flux, [30, 31] suggesting that ab initio-based approach that incorporates the free energy of the gas phase is feasible for investigating surface structures.

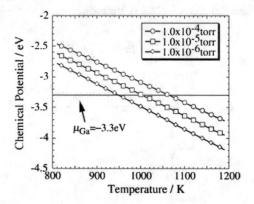

Fig. 4.3 Calculated gas phase chemical potential of Ga atom as a function of temperature for various BEP in [7]. Solid line denotes the value of adsorption energy of Ga adatom on GaAs(001)-(2×4) $\beta2$ surface (-3.3 eV). Reproduced with permission from Y. Kangawa, T. Ito, A. Taguchi, K. Shiraishi, and T. Ohachi, Surf. Sci. **493**, 178 (2001). Copyright (2001) by Elsevier

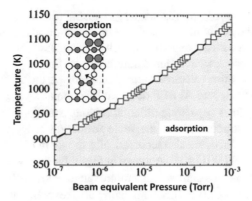

Fig. 4.4 Calculated surface phase diagram for Ga adatom on GaAs(001)-(2 × 4) β2 surface as functions of temperature and BEP in [7]. Geometry of GaAs(001)-(2 × 4) β2 surface is also shown, where filled and empty circles denote Ga and As atoms, respectively. The most stable adsorption site of Ga adatom is represent by an arrow Reproduced with permission from Y. Kangawa, T. Ito, A. Taguchi, K. Shiraishi, and T. Ohachi, Surf. Sci. **493**, 178 (2001). Copyright (2001) by Elsevier

4.2 Surface Reconstructions on III-Nitride Compounds

The surface energy calculations based on ab initio calculations for various surface structures of III-nitride compounds have revealed that the stable surface reconstructions are dependent on the chemical potential of constituent atomic species [5, 32–40]. Although these ab initio studies successfully elucidated some aspects of the surface-related issues, their results do not include growth parameters, such as BEP and temperature. By using surface phase diagrams, surface reconstructions in III-nitride compounds taking account of these parameters can be determined. Indeed, surface phase diagram calculations as functions of temperature and BEP have been accomplished for AlN, GaN, and InN surfaces with various orientations (See Fig. 4.1) [10–16]. In this section, the reconstructions on various III-nitride surfaces obtained by surface phase diagrams are discussed.

4.2.1 Polar AlN(0001) and (0001̄) Surfaces

Figure 4.5 displays the calculated surface formation energies for AlN polar surfaces shown in Fig. 4.1a as a function of Al chemical potential μ_{Al} by using (4.1) [34, 35, 36, 39, 40]. Here, the reconstructions considered are constructed on the basis of the electron counting (EC) rule, [41] in which dangling bonds of the topmost Al and N atoms are empty and filled by electrons, respectively. To satisfy the EC rule, the surface must be stabilized due to its semiconducting nature. In addition, the surfaces covered by Al atoms are also considered to determine the stability under Al-rich (high μ_{Al}) conditions. This energy diagram allows us to determine which reconstruction is the most stable. However, the reconstruction under growth conditions cannot be directly determined by this energy diagram. On the contrary, the surface diagram can be directly compared with the experiments because it is described as a function of the experimental parameters, such as temperature and BEP.

Figure 4.6a shows the calculated phase diagram of the AlN(0001) surface as functions of temperature and Al BEP obtained by comparing E_{ad} with μ_{gas} in (2.63) [23]. The boundary lines separating different regions correspond to temperature and BEP in which two structures have the same formation energy. The stable reconstructions on these surfaces are also schematically shown in Fig. 4.6a. The calculated phase diagram on AlN(0001) surface shown in Fig. 4.6a suggests that the Al metallic bilayer surface is stable in a narrow temperature range below 940 K at 1×10^{-8} Torr and below 1320 K at 1×10^{-2} Torr. This figure also reveals that the (2×2) surface with Al adatom is stable at 940–1030 K at 1×10^{-8} Torr and at 1320–1455 K at 1×10^{-2} Torr. The (2×2) surface with N adatom is favorable at lower Al BEP and higher temperatures because Al desorption is enhanced at these conditions. The Al bilayer surface is stabilized even though it does not satisfy the EC rule [41].

Figure 4.6b displays the calculated phase diagram on AlN(000$\bar{1}$) surface as functions of of temperature and Al BEP [23]. The (2×2) surface with Al adatom is stabilized below 1020 K at 1×10^{-8} Torr and below 1440 K at 1×10^{-2} Torr. However, the (1×1) surface with a monolayer of Al atoms (Al monolayer) is stable above 1020 K at 1×10^{-8} Torr and above 1440 K at 1×10^{-2} Torr. The surface phase diagram suggests that both surfaces can be obtained during the MBE growth of AlN for temperatures around 1200 K, and the Al monolayer (1×1) surface is favorable under Al-rich conditions.

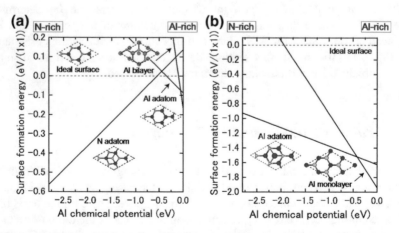

Fig. 4.5 Calculated surface formation energies for polar AlN surfaces with **a** (0001) and **b** (000$\bar{1}$) orientations as a function of Al chemical potential μ_{Al}. The origin of μ_{Al} is set the energy of bulk Al. Schematics of the surface structures under consideration are also shown. Large and small circles represent Al and N atoms, respectively

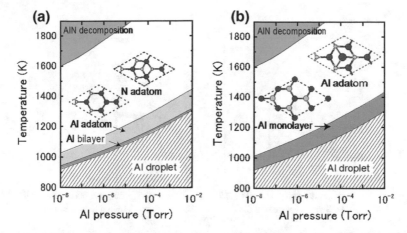

Fig. 4.6 Calculated phase diagrams for polar AlN surfaces with **a** (0001) and **b** (000$\bar{1}$) orientations as a function of temperature and Al BEP. Schematics of stable reconstructions on these surfaces are also shown. Notations of circles are the same as those in Fig. 4.5

4.2.2 Nonpolar AlN($1\bar{1}00$) and ($11\bar{2}0$) Surfaces

The epitaxial growth of nitride semiconductors has traditionally been performed along the polar [0001] direction, resulting in large polarization fields [42] along the growth direction. These fields reduce the radiative efficiency of quantum-well light emitters because they cause electron and hole separation. For optoelectronic device fabrication, there has been an increase in interest in the growth along nonpolar orientations, such as ($1\bar{1}00$) and ($11\bar{2}0$) planes, as shown in Fig. 4.1b and c, respectively [43]. Previous ab initio calculations have determined that the ideal surface is most stable over a large range of chemical potentials and that surfaces with Al adlayers are stabilized for Al-rich conditions [40]. Figure 4.7 shows the calculated formation energy of AlN nonpolar surfaces as a function of the Al chemical potential [23]. It is found that the ideal surfaces are stable over the wide range of growth conditions on both AlN($1\bar{1}00$) and ($11\bar{2}0$) surfaces. The calculated surface phase diagrams of AlN($1\bar{1}00$) and ($11\bar{2}0$) shown in Fig. 4.8 successfully reproduce the stability of the AlN nonpolar surface regardless of the growth conditions. The ideal surface appears over the entire temperature range. The Al bilayer and monolayer surfaces are always metastable. For the ideal surfaces, the N atoms relax outward, whereas the Al atoms relax inward and are accompanied by a charge transfer from the Al dangling bonds to the N dangling bonds. These ideal surfaces thus satisfy the EC rule [41] and are stabilized without any adsorption or desorption in the surface. It is thus concluded that the MBE growth of AlN on nonpolar orientations proceeds on the ideal surface over the entire range of Al BEP.

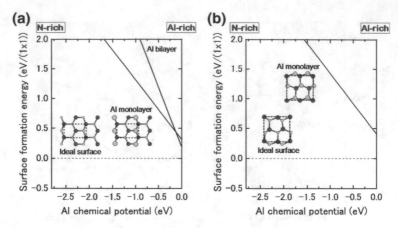

Fig. 4.7 Calculated surface formation energies for polar AlN surfaces with **a** $(1\bar{1}00)$ and **b** $(11\bar{2}0)$ orientations as a function of Al chemical potential μ_{Al}. The origin of μ_{Al} is set the energy of bulk Al. Schematics of the surface structures under consideration are also shown. Notations of circles are the same as those in Fig. 4.5

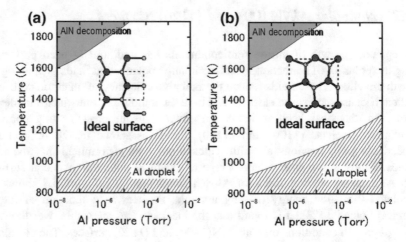

Fig. 4.8 Calculated phase diagrams for polar AlN surfaces with **a** $(1\bar{1}00)$ and **b** $(11\bar{2}0)$ orientations as functions of temperature and Al BEP. Schematics of stable reconstructions on these surfaces are also shown. Notations of circles are the same as those in Fig. 4.5

4.2.3 Semipolar AlN$(1\bar{1}01)$ and $(11\bar{2}0)$ Surfaces

In addition to nonpolar orientations, there is an increasing interest in crystal growth and device fabrication on semipolar orientations, such as $(1\bar{1}01)$ and $(11\bar{2}2)$, as shown in Fig. 4.1b and c, respectively, due to their reduced or negligible electric field [44–51]. Figure 4.9a displays the calculated surface formation energies of a

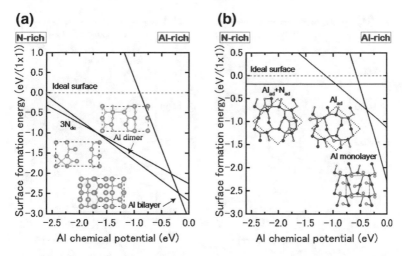

Fig. 4.9 Calculated surface formation energies for polar AlN surfaces with **a** $(1\bar{1}01)$ and **b** $(11\bar{2}2)$ orientations as a function of Al chemical potential μ_{Al}. The origin of μ_{Al} is set the energy of bulk Al. Schematics of the surface structures under consideration are also shown. The 2×2 unit cell for AlN$(1\bar{1}01)$ surface, and the 1×1 and c(2×2) unit cells for AlN$(11\bar{2}2)$ surface are shown by dashed rectangles. Notations of circles are the same as those in Fig. 4.5

semipolar AlN$(1\bar{1}01)$ surface, demonstrating that many types of surface structures are found depending on the Al chemical potential. The surfaces that have Al atoms at the topmost layer are stabilized over a wide range of Al chemical potentials. The phase diagram on semipolar AlN$(1\bar{1}01)$ surface is shown in Fig. 4.10a. With increasing temperature, the surface with metallic Al bilayer that is stabilized at low temperatures changes its structure into Al dimers. The metallic reconstruction is stabilized under Al-rich conditions similarly to the AlN(0001) surface. Therefore, many types of reconstructions could appear at approximately 1200 K (a typical MBE growth temperature) depending on the Al BEP. This conclusion suggests that AlN$(1\bar{1}01)$ surface growth kinetics depend on the growth temperatures.

The calculated surface formation energy for AlN$(11\bar{2}2)$ shown in Fig. 4.9b suggests that several reconstructions can occur depending on the Al chemical potential. The metallic reconstructions that have Al monolayer are stabilized under Al-rich conditions. However, the surface with Al and N adatoms is favored under N-rich conditions. Figure 4.10b shows the phase diagram of semipolar AlN$(11\bar{2}2)$ surface. The diagram suggests that the metallic reconstructions with Al monolayer emerge only at low temperatures and high Al-rich conditions. In contrast, the surface with Al and N adatoms is favored over a wide temperature range. The calculated phase diagram thus suggests that the growth kinetics on AlN$(11\bar{2}2)$ surface growth kinetics and its temperature dependence are similar to those on AlN$(1\bar{1}01)$ surface.

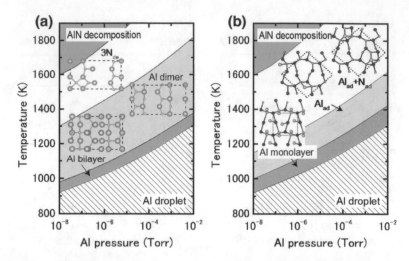

Fig. 4.10 Calculated phase diagrams for polar AlN surfaces with **a** $(1\bar{1}01)$ and **b** $(11\bar{2}2)$ orientations as functions of temperature and Al BEP. Schematics of stable reconstructions on these surfaces are also shown. Notations of circles are the same as those in Fig. 4.5

4.2.4 Polar GaN(0001) and (000\bar{1}) Surfaces

The reconstructed atomic structure during the MBE growth on the GaN(0001) surface under Ga-rich conditions has been the subject of many experimental and theoretical investigations. The (2×2) and pseudo-(1×1) surfaces have been observed on GaN(0001) under Ga-rich conditions by scanning tunneling microscopy (STM) [52, 53]. Furthermore, the coexistence of a "ghost" island with the (2×2)-like structure and a normal island with the pseudo-(1×1) structure has been found under excess Ga fluxes [54] There have been several ab initio theoretical studies for surface structures and adsorption behavior on these surfaces. The pseudo-(1×1) structure have proposed as the most stable state under the Ga-rich limit among various surface structures [5]. Although these ab initio studies have elucidated some aspects of the GaN surface, their results are limited to 0 K and did not incorporate growth parameters such as temperature and BEP.

Figure 4.11 depicts the calculated surface formation energy of GaN(0001) surfaces as a function of μ_{Ga} using (4.1) [5, 35, 38]. Here, the reconstructions considered are constructed on the basis of the EC rule [41]. To satisfy the EC rule, the surface must be stabilized due to its semiconducting nature. In addition, the surfaces covered by Ga atoms are also considered to determine the stability under Ga-rich (high μ_{Ga}) conditions. This energy diagram allows us to determine which reconstruction is the most stable. However, the reconstruction under growth conditions cannot be directly determined by this energy diagram. On the contrary, the surface diagram can be directly compared with the experiments because it is described as functions of the experimental parameters, such as temperature and Ga BEP.

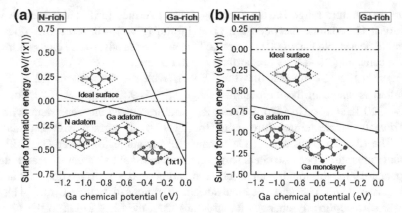

Fig. 4.11 Calculated surface formation energies for polar GaN surfaces with **a** (0001) and **b** (000$\bar{1}$) orientations as a function of Ga chemical potential μ_{Ga}. The origin of μ_{Ga} is set the energy of bulk Ga. Schematics of the surface structures under consideration are also shown. Large and small circles represent Ga and N atoms, respectively

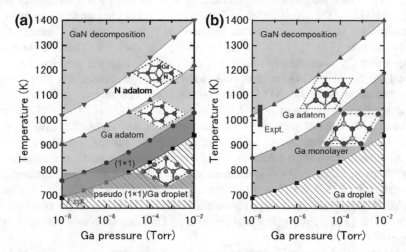

Fig. 4.12 Calculated phase diagrams for polar GaN surfaces with **a** (0001) and **b** (000$\bar{1}$) orientations as functions of temperature and Ga BEP. Shaded area denotes the temperature range during the growth in [54, 56]. Schematics of stable reconstructions on these surfaces are also shown. Notations of circles are the same as those in Fig. 4.11

Figure 4.12a shows the calculated phase diagram of the GaN(0001) surfaces as functions of temperature and Ga BEP [12, 14, 19]. The stable reconstructions on these surfaces are also schematically shown in Fig. 4.12. The pseudo–(1 × 1) surface is stable in the temperature range below 684 K at 1 × 10^{-8} Torr and below 943 K at 1 × 10^{-2} Torr. This stability is qualitatively consistent with the experimental stable temperature range for the pseudo–(1×1) surface [55]. The structure with additional Ga adatoms between the (1×1) and (2×2)-Ga structures does not appear to be a stable GaN(0001) structure because the Ga adsorption energy remains almost constant (2.6–2.8 eV) regardless of Ga coverage.

Figure 4.12a also reveals that the (2×2) with the Ga adatom [(2×2)-Ga, hereafter] is stable in the temperature range of 767–900 K at 1×10^{-8} Torr and 1028–1220 K at 1×10^{-2} Torr. These temperature ranges are consistent with the experimental stable temperature range for the (2×2) surface with Ga adatoms [57]. The (2×2) surface with Ga vacancy is favorable for lower Ga BEP and higher temperatures because Ga desorption is enhanced at lower Ga BEP and higher temperatures. In addition, the ideal (cleaved and unrelaxed) surface does not appear in the phase diagram because the ideal surface does not satisfy the EC rule [41]. The (2×2) surface directly changes its structure from the (2×2)-Ga to the (2×2) with Ga vacancy at lower Ga BEP and at higher temperatures.

From an experimental perspective, the (2×2) surface is often observed following an interruption in the Ga flux [56]. The phase diagram in Fig. 4.12a qualitatively agrees with this experimental finding because a decrease in Ga BEP prefers the (2×2)-Ga to the pseudo–(1×1) and (1×1) surfaces at a certain temperature (e.g., ~ 800 K). The shaded area in Fig. 4.12a denotes the temperature range for submonolayer GaN deposition. This temperature range includes the stable regions of the pseudo–(1×1), (1×1), and (2×2)–Ga surfaces. Thus, these results suggest that Ga adsorption or desorption can easily change the pseudo–(1×1) or (1×1) to the (2×2)-Ga surface and vice versa, depending on Ga BEP. This is also consistent with the STM observations [54]. The atomic structure of the reconstructions during and after MBE growth on the GaN(0001) surface under Ga-rich conditions has been studied by experimental and theoretical investigations [33]. The STM observations have clarified that the surface exhibits a (1×1) structure, and depositing additional Ga atoms onto this surface results in the (3×3), (6×6) and c(6×12) reconstructions. Based on ab initio calculations, it was determined that the (1×1) structure consists of a monolayer of Ga atoms bonded at the upper most sites above the topmost N atoms of an N-terminated bilayer. The (3×3) reconstruction consists of Ga adatoms bonded on top of Ga adlayer.

The calculated surface formation energy as a function of μ_{Ga} using (4.1) for GaN(000$\bar{1}$) surface is shown in Fig. 4.11b [23, 38]. The results suggest that the surfaces with Ga adatoms and a Ga monolayer can be stabilized. However, as mentioned previously, the reconstruction under growth conditions cannot be directly determined by the formation energy. Figure 4.12b shows the calculated phase diagram of the GaN(000$\bar{1}$) surfaces as functions of temperature and Ga BEP [23]. The (2×2) surface with Ga adatoms at hexagonal (H3) site is stabilized beyond 850 K at 1×10^{-8} Torr and below 1190 K at 1×10^{-2} Torr. On the other hand, the (1×1) surface that has a monolayer of Ga atoms is stable below 850 K at 1×10^{-8} Torr and 1190 K at 1×10^{-2} Torr. The surface phase diagram suggests that both surfaces can form at experimental temperatures around ∼ 1070 K, and the (1 × 1) surface with a monolayer of Ga atoms is favorable under Ga-rich conditions. Because the MBE on the GaN(000$\bar{1}$) surface has been performed under Ga-rich conditions, the calculated result is qualitatively consistent with the experimental stable temperature range for the (1 × 1) surface [56]. In addition, the ideal surface does not appear in the phase diagram because the ideal surface does not satisfy the EC rule [41].

4.2.5 Nonpolar GaN(1$\bar{1}$00) and (11$\bar{2}$0) Surfaces

The reconstructions on nonpolar GaN(1$\bar{1}$00) and (11$\bar{2}$0) surfaces are very simple, as shown in Figs. 4.13 and 4.14 [32, 38]. The calculated surface formation energies shown in Fig. 4.13 demonstrate that the ideal surface is stabilized over a wide Ga chemical potential range. In contrast, the calculated surface phase diagrams shown in Fig. 4.14 suggest that the ideal surface appears beyond the temperature range of 725–1030 K and 770–1100 K on GaN(1$\bar{1}$00) and (11$\bar{2}$0) surfaces, respectively [23]. However, the Ga adlayer surfaces are stable only at lower temperatures. For the ideal surfaces, the N atoms relax outward whereas the Ga atoms relax inward, which are accompanied by a charge transfer from the Ga dangling bonds to the N dangling bonds. As a result of this charge transfer, the ideal surfaces satisfy the EC rule [41] and are stabilized without any adsorption or desorption to the surface. Therefore, the MBE growth proceeds on the ideal GaN(11$\bar{2}$0) surface regardless of Ga BEP at the conventional growth temperatures.

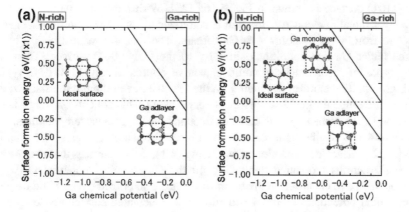

Fig. 4.13 Calculated surface formation energies for polar GaN surfaces with **a** (1$\bar{1}$00) and **b** (11$\bar{2}$0) orientations as a function of Ga chemical potential μ_{Ga}. The origin of μ_{Ga} is set the energy of bulk Ga. Schematics of the surface structures under consideration are also shown. Notations of circles are the same as those in Fig. 4.11

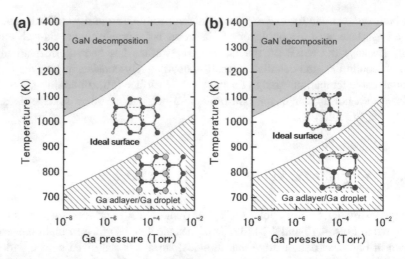

Fig. 4.14 Calculated phase diagrams for polar GaN surfaces with **a** $(1\bar{1}00)$ and **b** $(11\bar{2}0)$ orientations as functions of temperature and Ga BEP. Schematics of stable reconstructions on these surfaces are also shown. Notations of circles are the same as those in Fig. 4.11

4.2.6 Semipolar GaN$(1\bar{1}01)$ and $(11\bar{2}0)$ Surfaces

Figure 4.15a displays the calculated surface formation energies of semipolar GaN$(1\bar{1}01)$ surface as a function of μ_{Ga} using (4.1), demonstrating that many types of reconstructions are found depending on the Ga chemical potential [17]. The surfaces that have Ga atoms at the topmost layer are stabilized over a wide range of Ga chemical potentials. The calculated surface phase diagram on semipolar GaN$(1\bar{1}01)$ surface is shown in Fig. 4.16a [23]. With increasing temperature, the Ga bilayer metallic reconstruction that is stabilized at low temperatures changes its structure from a Ga monolayer to Ga dimers. The metallic reconstruction is stabilized under Ga-rich conditions similarly to the GaN(0001) surface. Therefore, many types of reconstructions could appear at approximately 1100 K (a typical MBE growth temperature) depending on the Ga BEP, even though the stabilization temperature range for the ideal surface is very narrow. This conclusion suggests that the GaN$(1\bar{1}01)$ surface growth kinetics depends on the growth temperatures.

Figure 4.15b displays the calculated surface formation energies of semipolar GaN$(11\bar{2}2)$ surface as a function of μ_{Ga} using (4.1), demonstrating that many types of reconstructions are found depending on the Ga chemical potential [13]. The surfaces that have Ga atoms at the topmost layer are stabilized over a wide range of Ga chemical potentials. Figure 4.16b shows the calculated surface phase diagram on semipolar GaN$(11\bar{2}2)$ surface [23]. The diagram suggests that the metallic reconstructions with a Ga adlayer or a monolayer emerge only at low temperatures and high Ga-rich conditions. In contrast, the Ga adatom surface is favored over a wide temperature range. The calculated surface phase diagram agrees well with the

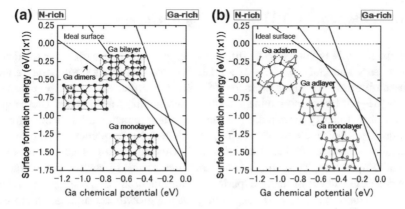

Fig. 4.15 Calculated surface formation energies for polar GaN surfaces with **a** $(1\bar{1}01)$ and **b** $(11\bar{2}2)$ orientations as a function of Ga chemical potential μ_{Ga}. The origin of μ_{Ga} is set the energy of bulk Ga. Schematics of the surface structures under consideration are also shown. Notations of circles are the same as those in Fig. 4.11

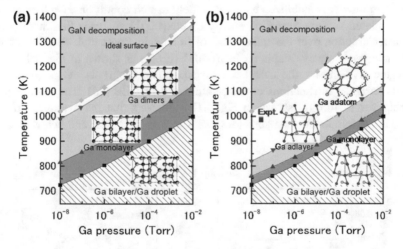

Fig. 4.16 Calculated phase diagrams for polar GaN surfaces with **a** $(1\bar{1}01)$ and **b** $(11\bar{2}2)$ orientations as functions of temperature and Ga BEP. Schematics of stable reconstructions on these surfaces are also shown. Notations of circles are the same as those in Fig. 4.11. Temperature range for the MBE growth in [58] is also shown

experimental results in which the Ga monolayer surface was formed under high Ga fluxes near the Ga accumulation (droplet) onset in the plasma-assisted MBE ($T \sim 1000$ K) [58]. The calculated phase diagram thus suggests that the growth kinetics and its temperature dependence on GaN$(11\bar{2}2)$ surface are similar to those on GaN$(1\bar{1}01)$ surface.

4.2.7 Polar InN(0001) and (000$\bar{1}$) Surfaces

High quality InN is known to be difficult to grow compared with other III-nitrides, such as AlN and GaN, because of a relatively low dissociation temperature and a high equilibrium N_2 vapor pressure [59, 60]. Nevertheless, several experimental studies [61–63] have reported the growth of high-quality InN crystals. These reports suggest that polarity termination is an important consideration when growing high-quality III-nitride compounds [64]. The InN radio frequency MBE growth on a sapphire substrate produced surfaces with varying polarities depending on the growth temperature. Thus, the atomic structures of the In surface reconstructions have been the subject of many experimental and theoretical investigations. Several stable structures on InN(0001) depending on the growth conditions, where an In monolayer directly above the surface N atoms is stabilized on a InN(0001) surface, have been proposed on the basis of surface formation energy [37, 38]. Figure 4.17 displays the calculated surface formation energy for InN(0001) and (000$\bar{1}$) surfaces as a function of μ_{In} using (4.1). There are three types of reconstructions including the ideal surface and depending on the In chemical potential. The metallic surface with an In bilayer is stabilized under extremely In-rich conditions. On the contrary, the calculated surface formation energy of InN(000$\bar{1}$) suggests that there is only one reconstruction over the entire range of In chemical potential. The surface with In monolayer is always stable on InN(000$\bar{1}$) surface as shown in Fig. 4.17b.

Figure 4.18 shows the calculated surface phase diagrams of InN(0001) and (000$\bar{1}$) surfaces as functions of temperature and In BEP [23]. The surface phase diagram for InN(0001) shown in Fig. 4.18a indicates that the metallic surface is stable in the temperature range below 695 K at 1×10^{-8} Torr and below 850 K at

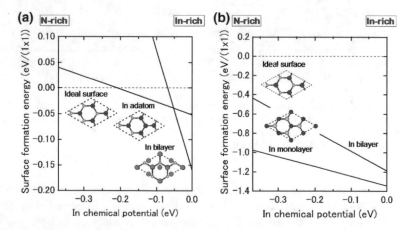

Fig. 4.17 Calculated surface formation energies for polar InN surfaces with **a** (0001) and **b** (000$\bar{1}$) orientations as a function of In chemical potential μ_{In}. The origin of μ_{In} is set the energy of bulk In. Schematics of the surface structures under consideration are also shown. Large and small circles represent In and N atoms, respectively

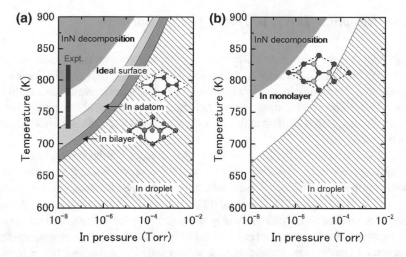

Fig. 4.18 Calculated phase diagrams for polar InN surfaces with **a** (0001) and **b** (000$\bar{1}$) orientations as functions of temperature and In BEP. Schematics of stable reconstructions on these surfaces are also shown. Notations of circles are the same as those in Fig. 4.17. Temperature range for the MBE growth in [61–63] is also shown

1×10^{-4} Torr. The (2×2) surface with an In adatom is stabilized under the moderate conditions of 695–725 K at 1×10^{-8} Torr and 860–890 K at 1×10^{-4} Torr. The ideal surface is favored for lower In BEP and higher temperatures because In desorption is enhanced at lower In BEP and higher temperatures. Therefore, during the MBE growth of InN (725–825 K), the surface changes from the metallic surface with In bilayer to the ideal surface via the (2×2) structure with In adatom at a lower In BEP and higher temperatures. Although experimental data are not available for comparison to the calculated results, trends obtained by theoretical calculations reasonably agrees with the formation energies obtained by previous ab initio calculations [37, 38]. Both the surface with In bilayer and the ideal surface are stabilized, even though they do not satisfy the electron EC rule [41]. These stable InN(0001) surface structures differ from those found for other closely related III-nitride compounds, such as AlN and GaN, over the entire temperature and pressure range.

The surface phase diagram on InN(000$\bar{1}$) shown in Fig. 4.18b reveals that In monolayer surface is always stable. The phase diagram also suggests that the In monolayer surface is stabilized regardless of the growth conditions, which is different from the results of GaN(000$\bar{1}$) surface. In the case of GaN surfaces, the lowest energy surface structure is a Ga monolayer and occurs only at low temperatures. However, under N-rich conditions, the surface with a Ga adatom on the H3 site in the (2×2) unit cell has the lowest energy, as shown in Fig. 4.11b. The stabilization of In monolayer on InN(000$\bar{1}$) surface could be related to the interatomic distances between the In monolayer (~ 3.16 Å) atoms, which are similar to those with bulk In (3.28 Å) with tetragonal symmetry.

4.2.8 Nonpolar InN(1$\bar{1}$00) and (11$\bar{2}$0) Surfaces

Figure 4.19 shows the calculated surface formation energies for nonpolar InN(1$\bar{1}$00) and (11$\bar{2}$0) surfaces as a function of μ_{In} using (4.1) [23, 38]. This figure suggests that the trends in stable structures in InN nonpolar planes are similar to those in GaN. The ideal surface is stable at moderate and high N/In ratios, and the metallic reconstruction similar to those in GaN surfaces [32, 38] is stabilized under In-rich conditions.

The calculated surface phase diagrams for nonpolar InN surfaces, shown in Fig. 4.20, successfully reproduce these structural characteristics depending on the growth conditions [23]. The ideal surface appears beyond the temperature range of 720–1010 K and 740–1035 K on InN(1$\bar{1}$00) and (11$\bar{2}$0) surfaces, respectively, whereas the surfaces with an In monolayer are stable at lower temperatures. For the ideal surfaces, the N atoms relax outward similar to the GaN nonpolar surfaces, whereas the In atoms relax inward and are accompanied by a charge transfer from the In dangling bonds to the N dangling bonds. The ideal surfaces thus satisfy the EC rule [41] and are stabilized without any adsorption or desorption from the surface. As expected from nonpolar orientations, the MBE growth of InN proceeds with over a wide range of In BEP and at the conventional growth temperatures (725–825 K).

4.2.9 Semipolar InN(1$\bar{1}$01) and (11$\bar{2}$0) Surfaces

Figure 4.21 shows the calculated surface formation energies for InN semipolar surfaces as a function of μ_{In} using (4.1). The stable surface structures shown in

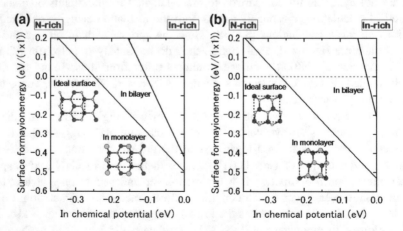

Fig. 4.19 Calculated surface formation energies for polar InN surfaces with **a** (1$\bar{1}$00) and **b** (11$\bar{2}$0) orientations as a function of In chemical potential μ_{In}. The origin of μ_{In} is set the energy of bulk In. Schematics of the surface structures under consideration are also shown. Notations of circles are the same as those in Fig. 4.17

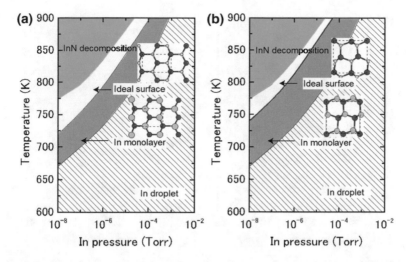

Fig. 4.20 Calculated phase diagrams for polar InN surfaces with **a** (1$\bar{1}$00) and **b** (11$\bar{2}$0) orientations as functions of temperature and In BEP. Schematics of stable reconstructions on these surfaces are also shown. Notations of circles are the same as those in Fig. 4.17

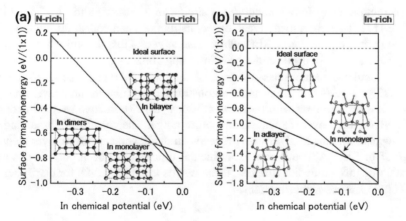

Fig. 4.21 Calculated surface formation energies for polar InN surfaces with **a** (1$\bar{1}$01) and **b** (11$\bar{2}$2) orientations as a function of In chemical potential μ_{In}. The origin of μ_{In} is set the energy of bulk In. Schematics of the surface structures under consideration are also shown. Notations of circles are the same as those in Fig. 4.17

Fig. 4.21 are slightly different from those on semipolar GaN surfaces shown in Fig. 4.15 because of the narrow chemical potential range of In [13, 16]. For the InN(1$\bar{1}$01) surface shown in Fig. 4.22a, there are several possible reconstructions depending on temperature and In BEP [23]. The metallic reconstruction with the In bilayer that is stabilized at low temperatures changes its In monolayer structure to contain In dimers upon higher temperatures. The stabilization of metallic

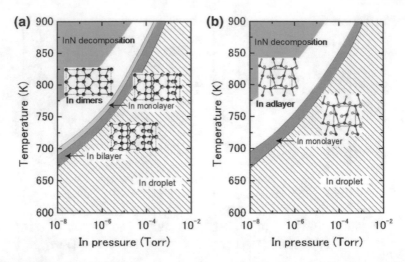

Fig. 4.22 Calculated phase diagrams for polar InN surfaces with **a** $(1\bar{1}01)$ and **b** $(11\bar{2}2)$ orientations as functions of temperature and In BEP. Schematics of stable reconstructions on these surfaces are also shown. Notations of circles are the same as those in Fig. 4.17

reconstruction under In-rich condition is similar to what occurs for the GaN($1\bar{1}01$) surface [17]. Therefore, many types of reconstructions can appear during the MBE growth depending on the In BEP, which also suggests that the growth kinetics on InN($1\bar{1}01$) surface depend on the growth temperatures.

The calculated surface phase diagram of semipolar InN($11\bar{2}2$) surface shown in Fig. 4.22b suggests that the metallic reconstruction with an In monolayer emerges only at low temperatures with In-rich conditions [13, 23]. In contrast, the In adlayer surface is favored over the wide temperature range. The calculated phase diagram thus suggests that the MBE growth proceeds on the In adlayer surface over the wide range of In BEP. Although few experimental reports have been published for the structure of InN semipolar surfaces, these results are informative for clarifying the InN($11\bar{2}2$) surface reconstructions.

4.3 Hydrogen Adsorption on III-Nitride Compounds

During typical vapor-phase epitaxy such as MOVPE, the surface interacts with H-rich ambient. When we consider high H_2 pressure conditions corresponding to MOVPE growth, the stable structures differ from those at low H_2 pressure. Thus, clarifying the reconstruction and taking the hydrogen adsorption into account are indispensable when investigating the stability of III-nitride surfaces during the MOVPE growth. In this section, the stability of hydrogen, such as H_2 and NH_2, on various III-nitride surfaces, is systematically investigated on the basis of surface phase diagrams, in which gas-phase chemical potentials of H_2 is taken into account.

4.3.1 Structures of AlN Surfaces with Hydrogen

Figure 4.23 shows the diagrams of stable AlN surfaces as functions of μ_{Al} and μ_H using (4.1) [20, 21, 24, 39]. The boundary lines separating different regions correspond to chemical potentials for which two structures have the same formation energy. For polar AlN(0001) surface, as shown in Fig. 4.23a, when $\mu_H - \mu_{H_2} \approx 0$ eV (μ_{H_2} is the chemical potential of H_2 molecule), the (0001) surface with NH_3 and NH_2 (Al–NH_3 + 3Al–NH_2) is favorable under N-rich conditions while that with NH_3 and three Al–H bonds (Al–NH_3 + 3Al–H) is stabilized under Al-rich conditions. For $\mu_H - \mu_{H_2} \leq -1.4$ eV, the (0001) surface with an N adatom (N_{ad}) and that with an Al adatom (Al_{ad}) are stabilized under N-rich and Al-rich conditions, respectively. The surface with Al bilayer is stabilized under extreme Al-rich conditions satisfying $\mu_{Al} \geq -0.04$ eV. However, this structure is not stable during the growth since H_2 pressures of 76 Torr (0.1 atm) at 1300–1800 K, which is a typical MOVPE condition, correspond to $\mu_H - \mu_{H_2}$ ranging from -1.6 to -1.1 eV. For $-1.6 \leq \mu_H - \mu_{H_2} \leq -1.1$ eV, the (0001) surface tends to form Al–H bonds (3Al–H) in under Al–rich conditions, whereas hydrogen terminated N adatoms such as NH and NH_2 (N_{ad}–H + Al–H and N_{ad}–H + Al–NH_2) are stabilized under N–rich conditions. It is thus expected that the reconstructions with H atoms as well as the surface with an N adatom emerge under H-rich ambient.

The reconstructions with N adatoms are not favorable on AlN(000$\bar{1}$) surface, as shown in Fig. 4.23b. This is because the formation of N–N bonds leads to nitrogen desorption as N_2 molecules. The (000$\bar{1}$) surface forms N–H bonds (3N–H) under extreme H-rich conditions while Al adsorption occurs under H-poor conditions. The surface with Al monolayer and that with an Al adatom at the tetrahedral (T4) site are stabilized under Al- rich and N-rich conditions, respectively. The transition point between adlayer and adatom is located at $\mu_{Al} = -0.42$ eV. If we consider H-rich conditions during the MOVPE growth ($-1.6 \leq \mu_H - \mu_{H_2} \leq -1.1$ eV), these reconstructions can emerge depending on temperature and Al pressure.

For nonpolar AlN(1$\bar{1}$00) surface shown in Fig. 4.23c, when $\mu_H - \mu_{H_2} \approx 0$ eV, the (1$\bar{1}$00) surface with H atoms and NH_2 molecules (2N–H + 2Al–NH_2) that terminate the outermost N and Al atoms, respectively, is favorable under N-rich conditions. On the other hand, the H-terminated (1$\bar{1}$00) surface (2N–H + 2Al–H) is stabilized under Al-rich conditions. The intermediate surface structure (2N–H + Al–H + Al–NH_2) between 2N–H + 2Al–NH_2 and 2N–H + 2Al–H is also stabilized for the (relaxed) ideal surface consisting of the outermost Al and N atoms is stabilized irrespective of the chemical potential of Al. Even under Al-rich conditions, the surfaces with Al adatoms such as Al monolayer or bilayers cannot be stabilized. Since H_2 pressures of 76 Torr (0.1 atm) at 1300–1800 K, which is a typical MOVPE condition, correspond to $\mu_H - \mu_{H_2}$ ranging from -1.6 to -1.1 eV, the ideal surface found to be stable during the growth. It is thus expected that the ideal (1$\bar{1}$00) surface without H atoms emerges even under H-rich ambient.

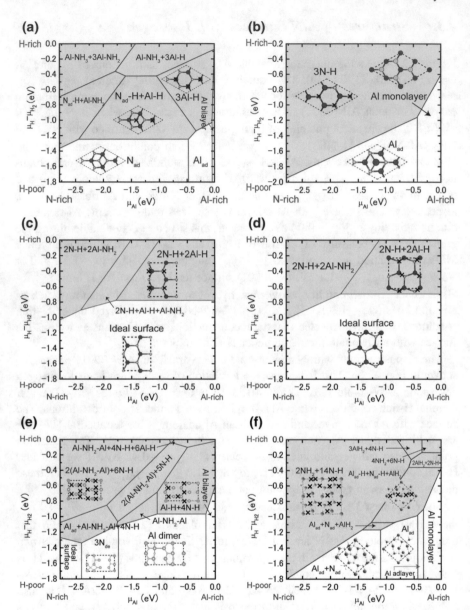

Fig. 4.23 Stable structures of polar **a** AlN(0001) and **b** AlN(000$\bar{1}$) surfaces, nonpolar **c** AlN(1$\bar{1}$00) and **d** AlN(11$\bar{2}$0) surfaces, and semipolar **e** AlN(1$\bar{1}$01) and **f** AlN(11$\bar{2}$2) surfaces as functions of μ_{Al} and μ_H. The origins of μ_H and μ_{Al} correspond to H_2 molecules and bulk Al (Al droplet) at $T = 0$ K, respectively. Stable region of the surfaces with hydrogen is emphasized by shaded area. Top views of surface structures are also shown. Notations of circles are the same as those in Fig. 4.5. The positions of H atoms in the H-terminated surfaces are marked by crosses

Figure 4.23d shows the diagram of stable $AlN(11\bar{2}0)$ surfaces as functions of μ_{Al} and μ_H. Similar to the $(1\bar{1}00)$ plane, for $\mu_H - \mu_{H_2} \approx 0$ eV the $(11\bar{2}0)$ surface forms both N–H and Al–NH$_2$ bonds (2N–H + 2Al–NH$_2$) under N-rich conditions while the surface with N–H and Al–H bonds (2N–H + 2Al–H) is favorable under Al-rich conditions. In contrast to the $(1\bar{1}00)$ plane, the surface with both Al–H and Al–NH$_2$ bonds (2N–H + Al–H + Al–NH$_2$) is always metastable on the $(11\bar{2}0)$ surface. For $\mu_H - \mu_{H_2} \leq -0.70$ eV, the ideal surface is stabilized irrespective of Al chemical potential. The surfaces with Al adatoms on the $(11\bar{2}0)$ plane are not stable even under Al-rich conditions. The calculated surface formation energies thus suggest that the surface with H atoms is metastable during the growth on the $(11\bar{2}0)$ plane. The ideal $(11\bar{2}0)$ surface without H atoms emerges even under H-rich ambient.

Figure 4.23e shows the diagram of stable $AlN(1\bar{1}01)$ surface as functions of μ_{Al} and μ_H. When $\mu_H - \mu_{H_2} \approx 0$ eV, the $(1\bar{1}01)$ surface with H atoms and NH$_2$ molecules (2(Al–NH$_2$)–Al + 6N–H) that terminate the outermost N and Al atoms, respectively, is favorable under N-rich conditions, whereas the H-terminated $(1\bar{1}01)$ surface with Al–H and N–H bonds including N desorption (Al–H + 4N–H) is stabilized under Al-rich conditions. Even under Al-rich conditions, the surfaces with Al adatoms, such as Al monolayer or bilayers cannot be stabilized. Since H$_2$ pressures of 76 Torr (0.1 atm) at 1300–1800 K, which is a typical MOVPE condition, correspond to $\mu_H - \mu_{H_2}$ ranging from -1.6 to -1.1 eV, the H-incorporated surface is not stabilized during the MOVPE growth.

For $AlN(11\bar{2}2)$ surface shown in Fig. 4.23f, when $\mu_H - \mu_{H_2}$ is close to 0 eV, which corresponds to extreme H-rich conditions, the (2×2) surface with NH$_2$ (2NH$_2$+ 14N–H) is favorable under N-rich conditions and the c(2×2) surface with N–H bonds (3AlH$_2$ + 4N–H) is stabilized under Al-rich conditions. For low $\mu_H - \mu_{H_2}$ with $-2.9 \leq \mu_{Al} \leq -1.2$ ($-1.2 \leq \mu_{Al} \leq -0.5$) eV, the c$(2 \times 2)$ surfaces with Al and N (Al) adatoms is stabilized under N-rich (Al-rich) conditions. The surface with the Al monolayer is stabilized under extreme Al-rich conditions satisfying $\mu_{Al} \geq -0.3$ eV. However, the surfaces covered by Al atoms are not stable during the MOVPE growth of the $AlN(11\bar{2}2)$ surface, since the MOVPE growth is performed under N-rich conditions. For moderately H–rich conditions, the surface tends to form Al–H and AlH$_2$ bonds (Al$_{ad}$ + N$_{ad}$ + AlH$_2$ and Al$_{ad}$–H + N$_{ad}$–H + AlH$_2$) in addition to the Al adatom. It is thus expected that the reconstructions with H atoms and the surface with Al adatoms will emerge under H-rich ambient. We also note that the stable surface structures under H-poor conditions are consistent with those obtained by previous calculations without taking account of H atoms in Sect. 4.2.3.

4.3.2   Surface Phase Diagrams for Hydrogen Adsorption on AlN Surfaces

Figure 4.24 shows calculated surface phase diagrams that exhibit a different surface phase diagram trend compared with those without H atoms shown in Figs. 4.6, 4.8, and 4.10 [20, 21, 24, 25]. Here, the surface phase diagrams are obtained assuming the H_2 pressure ($p_{H_2} = 76$ Torr) corresponds to H-rich conditions. The H-terminated surfaces with NH and NH_2 are primarily found over a wide range of temperatures and Al BEP. The calculated surface phase diagram for AlN(0001) surface shown in Fig. 4.24a demonstrates that the reconstructions with H atoms, such as 3Al–H and N_{ad}–H + Al–H, emerge below 1520 K [20]. However, the surface without H atoms (N_{ad}) can be formed above 1520 K even under H-rich conditions. This result suggests that there are several surface reconstructions, and the growth processes may change drastically depending on temperature and Al pressure. Because the N dangling bonds in N_{ad} are chemically active compared with the N–H and Al–H bonds in 3Al–H and N_{ad}–H + Al–H, the adsorption at high temperatures may be more efficient than at low temperatures. Furthermore, due to the presence of N_{ad} over a wide range of growth conditions at low H_2 pressures, the growth rate at low H_2 pressures is expected to be higher than at high H_2 pressures. For the AlN(000$\bar{1}$) surface, the N–H bonds (3N–H) are stabilized over a wide range of temperatures and Al pressures, as shown in Fig. 4.24b, which suggests that the growth processes on AlN(000$\bar{1}$) surface are insensitive to the growth conditions.

The calculated surface phase diagram for AlN(1$\bar{1}$00) surface shown in Fig. 4.24c demonstrates that the reconstruction with H atoms, 2 N–H + 2Al–H, emerges below 1060 K, whereas the ideal surface without H atoms can be observed only above 1060 K [21]. The surface phase diagram for AlN(11$\bar{2}$0) surface shown in Fig. 4.24c also demonstrate that the 2N–H + 2Al–H appears in a very narrow temperature range (below 990 K), and the ideal surface forms over a wide range of growth temperatures and pressures. These results imply that during growth the AlN nonpolar surfaces always form the ideal surfaces, even under H-rich conditions. The growth processes on nonpolar orientations are expected to be unchanged by temperatures and Al pressures. In contrast to the other nitrides, such as GaN and InN, the growth temperatures for AlN are too high to stabilize H atoms on nonpolar surfaces. The appearance of ideal surface occurs upon growth due to H atom desorption. By comparing the AlN nonpolar surface phase diagrams with polar orientations, as shown in Fig. 4.24a and b, it is expected that the AlN growth processes on nonpolar orientations are different from those on polar orientations.

In contrast, the surface phase diagram on AlN(1$\bar{1}$01) surface shown in Fig. 4.24e demonstrates that the surface with Al dimer is considerably stable over the wide range of temperatures and Al pressures: The other H-incorporated surfaces are found to be always metastable. This suggests that the reconstructions on AlN(1$\bar{1}$01) surface are also insensitive to hydrogen ambient at typical growth temperatures ranging from 1300 to 1800 K. The surface with Al dimer satisfies the

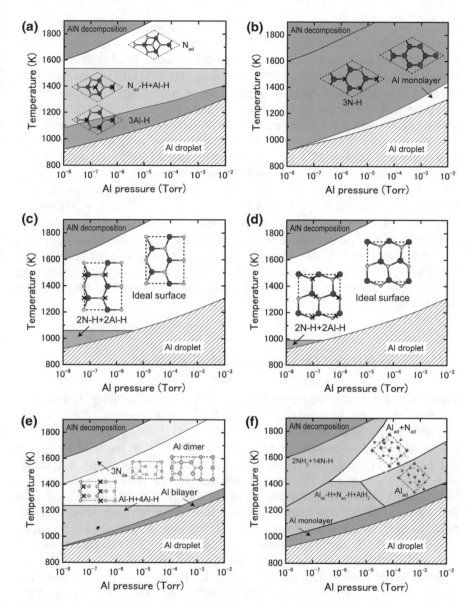

Fig. 4.24 Calculated surface phase diagram for H-adsorption on polar **a** AlN(0001) and **b** AlN(000$\bar{1}$) surfaces, nonpolar **c** AlN(1$\bar{1}$00) and **d** AlN(11$\bar{2}$0) surfaces, and semipolar **e** AlN(1$\bar{1}$01) and **f** AlN(11$\bar{2}$2) surfaces as functions of temperature and Al BEP under high H_2 pressure ($p_{H_2} = 76$ Torr) conditions. Top views of surface structures are also shown. Notations of circles are the same as those in Fig. 4.5. The positions of H atoms in the H-terminated surfaces are marked by crosses

EC rule [41]. Although the $(1\bar{1}01)$ orientation has N-polar, the ideal surface has two-coordinated N atoms. These two-coordinated N atoms on the ideal surface tends to desorb from the surface, resulting in the formation of Al dimers. It should be noted that the surface reconstructions on semipolar AlN$(1\bar{1}01)$ surface are quite different from those on GaN$(1\bar{1}01)$ surface, which is discussed in Sec. 4.3.4.

Figure 4.24f shows the calculated surface phase diagram of AlN$(11\bar{2}2)$ surface, which demonstrates that there are several reconstructions depending on temperature and Al pressure. The reconstructions with H atoms, such as Al$_{ad}$-H + N$_{ad}$–H + AlH$_2$ and 2NH$_2$ + 14N–H, emerge under low Al pressure conditions. In contrast, the c(2 × 2) surfaces without H atoms such as Al$_{ad}$ and Al$_{ad}$ + N$_{ad}$ are stabilized under high Al pressure conditions. Around 1400 K, which corresponds to the temperature of the MOVPE growth, the Al$_{ad}$ appears under low Al pressure conditions below 1×10^{-5} Torr, whereas the Al$_{ad}$–H + N$_{ad}$–H + AlH$_2$ is stabilized under high Al pressure conditions beyond 1×10^{-5} Torr. It is thus expected that the structural change from Al$_{ad}$ to Al$_{ad}$–H + N$_{ad}$–H + AlH$_2$ with Al pressure will cause the difference in the growth processes. The stability of these surface structures can be interpreted in terms of the adsorption energy of hydrogen (-1.1 eV/atom) and the EC rule [41].

4.3.3 Structures of GaN Surfaces with Hydrogen

Figure 4.25 shows the diagrams of stable GaN surfaces including hydrogen as functions of μ_{Ga} and μ_H using (4.1) [17, 18, 19, 27]. For polar GaN(0001) surface, as shown in Fig. 4.25a, the surfaces with H atoms are stabilized under H-rich (high μ_H) conditions. If we assume the pressures of H$_2$ and Ga as $p_{H_2} = 76$ and $p_{Ga} = 5.0 \times 10^{-4}$ Torr for 1270–1370 K, respectively [open circles in Fig. 4.25a], the surface with a topmost Ga atom terminated by an NH$_2$ molecule and an H-terminated N adatom attached to the other topmost Ga (N$_{ad}$–H + Ga-NH$_2$) is favored under N-rich conditions $\mu_{Ga} \leq -1.16$ eV. For relatively Ga-rich conditions, on the other hand, the H-terminated surface with an H-terminated N adatom (N$_{ad}$–H + Ga–H) is stabilized. Therefore, these surfaces are expected to emerge during the MOVPE depending on the growth conditions. Since both the N$_{ad}$–H + Ga–H and N$_{ad}$–H + Ga–NH$_2$ satisfy the EC rule, [41] the stabilization of the N$_{ad}$–H + Ga–NH$_2$ under N-rich conditions can be interpreted in terms of the desorption of Ga atoms. The topmost Ga atoms in the N$_{ad}$–H + Ga–H desorb and N atoms appear with decreasing μ_{Ga}. Owing to H$_2$ ambient under H-rich conditions, H atoms terminate the remaining N atoms, resulting in the formation of H-terminated N adatoms and Ga–NH$_2$ bonds.

The reconstructions with N adatoms are not favorable on GaN$(000\bar{1})$ surface, as shown in Fig. 4.25b. As mentioned before, this is because the formation of N–N bonds leads to nitrogen desorption as N$_2$ molecules. The $(000\bar{1})$ surface forms N–H bonds (3N–H) regardless of H chemical potential. The surface with Ga monolayer

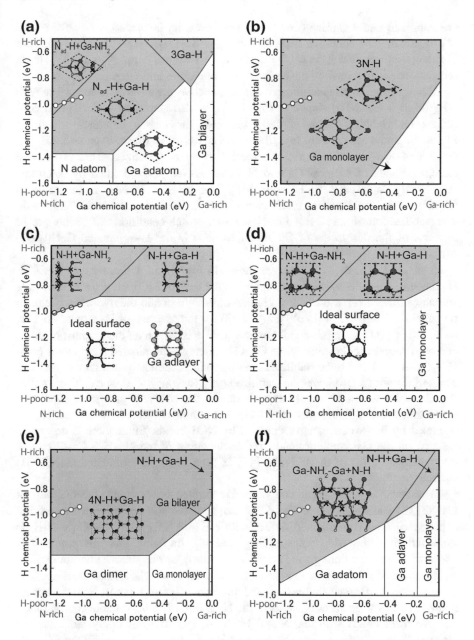

Fig. 4.25 Stable structures of polar **a** GaN(0001) and **b** GaN(0001̄) surfaces, nonpolar **c** GaN(11̄00) and **d** GaN(112̄0) surfaces, and semipolar **e** GaN(11̄01) and **f** GaN(112̄2) surfaces as functions of μ_{Ga} and μ_{H}. The origins of μ_{H} and μ_{Ga} correspond to H_2 molecules and bulk Ga (Ga droplet) at $T = 0$ K, respectively. Stable region of the surfaces with hydrogen is emphasized by shaded area. Stable region of the surfaces with hydrogen is emphasized by shaded area. Open circles indicate μ_{H} and μ_{Ga} with $p_{H_2} = 76$ and $p_{Ga} = 5 \times 10^{-4}$ Torr, respectively, ranging from 1270 to 1370 K. Top views of surface structures are also shown. Notations of circles are the same as those in Fig. 4.11. The positions of H atoms in the H-terminated surfaces are marked by crosses

are stabilized under Ga-rich conditions. The boundary between Ga monolayer and 3N–H satisfies the condition expressed as $\mu_H = 1.31\mu_H - 0.84$ eV. If we consider H-rich conditions during the MOVPE growth ($-1.6 \leq \mu_H - \mu_{H_2} \leq -1.1$ eV), these reconstructions can emerge depending on temperature and Ga pressure.

Although similar NH_2-terminated surface, such as the H-terminated surface with NH_2 (N–H + Ga–NH_2) shown in Fig. 4.25a, is also stabilized under H-rich conditions on GaN($1\bar{1}00$) and ($11\bar{2}0$) surfaces, the stable structures shown in Fig. 4.25c and d are different from those on GaN(0001) surface. Even under H-rich conditions $\mu_H - \mu_{H_2} \approx -1$ eV, the ideal surface is stabilized. This is because dangling bonds of topmost Ga and N atoms are empty and filled by electrons, respectively, satisfying the EC rule [41]. The N–H + Ga–NH_2 is favorable in addition to the ideal surface under N-rich conditions during the MOVPE growth. The stabilization of the N–H + Ga–NH_2 under N-rich condition can be interpreted in terms of the desorption of Ga atoms, as seen in the H-terminated GaN(0001) surface.

In the case of GaN($1\bar{1}01$) surface shown in Fig. 4.25e, when we consider high H_2 pressure conditions which correspond to the case of MOVPE growth, the stable structures differ from those in low H_2 pressures. We obtain the hydrogen chemical potential $\mu_H - \mu_{H_2} = -1.05$ eV using (2.63) for the pressure of H_2 around 76 Torr (0.1 atm) at 1300 K. In this condition, the surface in which all the topmost N atoms and one of topmost Ga atoms in the Ga–Ga dimers are terminated by H atoms (4N–H + Ga–H) is found to be stabilized over the wide chemical potential range of Ga. The surface with a Ga bilayer are stabilized only for $\mu_{Ga} \geq -0.04$ eV. This implies that the 4N–H + Ga–H usually appears during the MOVPE growth. It should be noted that the bonding states of Ga–Ga dimer and Ga–H bonds are completely occupied by the excess electrons caused by N–H bonds. Since there is no excess electron on the Ga dangling bonds, the stabilization of the 4N–H + Ga–H can be interpreted in terms of the EC rule, [41] as seen in the H-terminated GaN(0001) surface.

In contrast to these polar and nonpolar surface orientations, the diagram of GaN($11\bar{2}2$) surface shown in Fig. 4.25f manifests the absence of growth-condition dependence. The H-terminated surface with NH and NH_2 (Ga–NH_2–Ga + N-H) is stabilized over the wide range of μ_{Ga} and μ_H, implying that this structure is expected to emerge during the MOVPE regardless of the growth conditions. The stabilization of this surface is related to the polarity of GaN($11\bar{2}2$) surface. The N-terminated surface where the two- and three-coordinated topmost N atoms appear is the ideal cleavage surface. Because the N–H bond is a very stable configuration and can form bonds with Ga, N and H atoms, the H atoms easily terminate the topmost N atoms with a large increase in energy (~ 4 eV). To satisfy the EC rule, [41] three of the eight top N atoms have lone pairs. This structure is similar to the stable H-terminated GaN($000\bar{1}$) surface, which has a strong hydrogen affinity.

4.3.4 Surface Phase Diagrams for Hydrogen Adsorption on GaN Surfaces

Figure 4.26 displays the calculated surface phase diagrams of H-adsorbed GaN surface phase diagrams for polar (0001), polar (000$\bar{1}$), nonpolar (1$\bar{1}$00) nonpolar (11$\bar{2}$0), semipolar (1$\bar{1}$01) and semipolar (11$\bar{2}$2) orientations as functions of temperature and Ga BEP [17–19]. The surface phase diagrams are obtained assuming the H$_2$ pressure, p_{H_2} = 76 Torr (0.1 atm), corresponds to H-rich conditions. The adsorption of H atoms exhibits a different surface phase diagram trend compared with those without H atoms, as shown in Figs. 4.12, 4.14, and 4.16. The H-terminated surfaces with NH and NH$_2$ are typically formed over a wide range of temperatures and Ga BEP.

The surface phase diagram for polar GaN(0001) surface shown in Fig. 4.26a demonstrates that the N$_{ad}$–H + Ga–H can be formed from 1270 to 1370 K at $p_{Ga} \geq 1 \times 10^{-3}$ Torr [18]. The diagram also indicates that N$_{ad}$–H + Ga–H and N$_{ad}$–H + Ga–NH$_2$ are stabilized at low temperatures and high temperatures, respectively. Since both N$_{ad}$–H + Ga–H and N$_{ad}$–H + Ga–NH$_2$ satisfy the EC rule, [41] the stabilization of N$_{ad}$–H + Ga–NH$_2$ under N-rich conditions can be interpreted in terms of the desorption of Ga atoms. In N$_{ad}$–H + Ga–H, the topmost Ga atoms desorb and N atoms appear with decreasing μ_{Ga}. When the surface is exposed to H-rich conditions, H atoms immediately terminate the remaining N atoms, resulting in the formation of H-terminated N adatoms and Ga–NH$_2$ bonds.

For GaN(000$\bar{1}$) surfaces shown in Fig. 4.26b, the surfaces terminated by H atoms (3N–H) is stabilized over the entire temperature and Ga BEP. This suggests that the reconstructions on GaN(000$\bar{1}$) surface is insensitive to the growth conditions. The reconstructions with Ga adatoms, that are stabilized without H atoms shown in Fig. 4.12b, are not favorable due to the N–N bond formation which leads to the desorption of N atoms as N$_2$ molecules. The N–H bond has a very stable configuration among various bonds between Ga, N and H atoms, so that the H atoms easily terminate the topmost N atoms and a large energy gain ($\sim$4 eV) occurs. The formation of three N–H bonds leads to a charge transfer from the N–H bond to the remaining N dangling bond, which results in the formation of filled dangling bonds (lone pairs) to satisfy the EC rule [41]. This structure corresponds to the strong affinity of hydrogen.

The nonpolar GaN(1$\bar{1}$00) and (1$\bar{1}$20) surfaces shown in Fig. 4.26c and d, respectively, form NH$_2$-terminated surfaces similar to the GaN(0001) surface [17, 19]. However, the stable structures are different from those found on the GaN(0001) surface, depending on the growth temperature. The ideal surface shown in Fig. 4.14 is stable even under the H-rich conditions at 1200–1400 K at $p_{Ga} \geq 1 \times 10^{-4}$ Torr. This is because dangling bonds of the topmost Ga are empty, and the topmost N atoms are filled by electrons, which both satisfy the EC rule [41]. The N–H + Ga–NH$_2$ is favorable below 1300 K for $p_{Ga} \leq 1 \times 10^{-4}$ Torr. Therefore, two different types of reconstructions can occur in the MOVPE growth on nonpolar orientations.

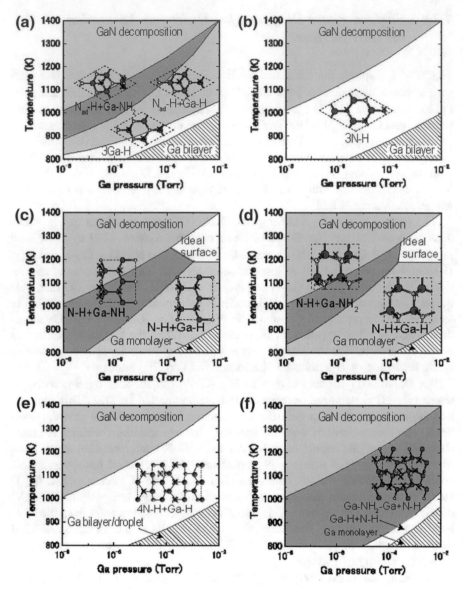

Fig. 4.26 Calculated surface phase diagram for H-adsorption on polar **a** GaN(0001) and **b** GaN(000$\bar{1}$) surfaces, nonpolar **c** GaN(1$\bar{1}$00) and **d** GaN(11$\bar{2}$0) surfaces, and semipolar **e** GaN(1$\bar{1}$01) and **f** GaN(11$\bar{2}$2) surfaces as functions of temperature and Ga BEP under high H_2 pressure ($p_{H_2} = 76$ Torr) conditions. Top views of surface structures are also shown. Notations of circles are the same as those in Fig. 4.11. The positions of H atoms in the H-terminated surfaces are marked by crosses

In contrast to polar and nonpolar surfaces, the surface phase diagram for semipolar GaN($1\bar{1}01$) surface is simple, as shown in Fig. 4.26e [17]. The surface with N atoms at the top layer and also Ga atoms at the top layer with Ga–Ga dimers that are terminated by H atoms is stable over a wide range of Ga BEP and temperatures, suggesting that the 4N–H + Ga–H usually appears during the MOVPE growth. The bonding states of the Ga–Ga dimer and Ga–H bonds are completely occupied by the excess electrons due to the N–H bonds. Thus, the stability of 4N–H + Ga–H can be interpreted in terms of the EC rule, [41] as seen in the H-adsorbed polar and nonpolar GaN surfaces. The surface phase diagram of GaN($11\bar{2}2$) surface shown in Fig. 4.26f indicates that the stable region of the Ga–NH$_2$–Ga + N–H expands over the wide temperature and Ga BEP range, suggesting that this structure always emerges at temperatures ranging from 1200 to 1400 K regardless of Ga pressure [18]. This suggests that Ga–NH$_2$–Ga + N–H emerges during the MOVPE regardless of the growth conditions. The stabilization is related to the polarity of GaN($11\bar{2}2$) surface: The N-terminated surface where the two- and three-coordinated topmost N atoms appear is the ideal cleavage surface. Because the N–H bond is a very stable configuration and can form bonds with Ga, N and H atoms, the H atoms easily terminate the topmost N atoms. To satisfy the EC rule, [41] three of the eight top N atoms have lone pairs. This structure is similar to the stable H-terminated GaN($000\bar{1}$) surface, which has a strong hydrogen affinity.

4.3.5 Structures of InN Surfaces with Hydrogen

It is known that growth on InN is prevented with increasing H$_2$ pressure and when N$_2$ is used as the carrier gas. Thermodynamic analysis has also shown that the InN deposition rate decreases with increasing hydrogen pressure, [65] suggesting that surface reconstructions and growth kinetics on InN surfaces are different than on AlN and GaN surfaces. From a theoretical perspective, the reconstructions on nonpolar and semipolar InN surfaces under different MBE growth conditions have been investigated, and several stable structures have been found depending on the growth conditions [16, 37, 38]. However, the stability and its temperature and pressure dependence on InN surfaces have been less pursued than for clean InN surfaces.

Figure 4.27 shows the diagrams of stable InN surfaces including hydrogen as functions of μ_{In} and μ_{H} using (4.1) [18]. For polar InN(0001) and ($000\bar{1}$) surfaces shown in Fig. 4.27a and b, respectively, the surface with In adatom and that with In bilayer are stabilized at moderate In/N rations and In-rich conditions with H-poor (low μ_H) conditions, respectively. For H-rich (high μ_H) conditions, the H-terminated surfaces with N–H and NH$_2$ (N$_{ad}$–H + In–NH$_2$ for InN(0001) surface and 4N–H for InN($000\bar{1}$) surface) are stabilized. If we assume the pressures of H$_2$ and In as $p_{H_2} = 76$ and $p_{In} = 5.0 \times 10^{-4}$ Torr for $770 - 900$ K, respectively, [open circles in Fig. 4.27a and b], these surface are favorable over the wide range of μ_H and μ_{In}. Therefore, the H-terminated surfaces, such as N$_{ad}$–H + In–NH$_2$ on InN (0001) surface [Fig. 4.27a] and the 4N–H on InN(0001) [Fig. 4.27b] is expected to emerge during the MOVPE regardless of the growth conditions.

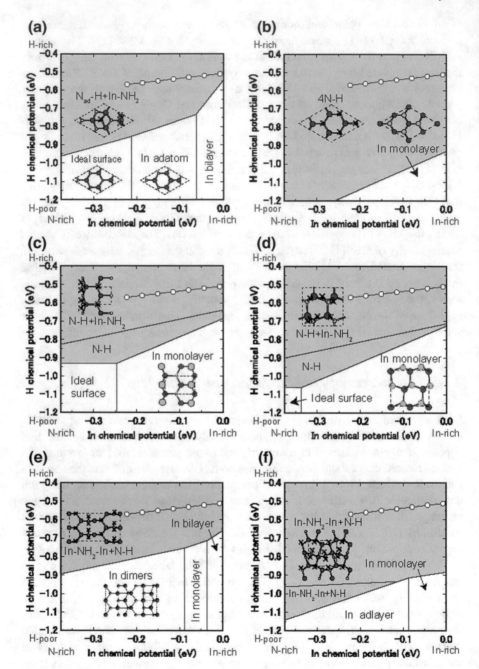

◀**Fig. 4.27** Stable structures of polar **a** InN(0001) and **b** InN(000$\bar{1}$) surfaces, nonpolar **c** InN(1$\bar{1}$00) and **d** InN(11$\bar{2}$0) surfaces, and semipolar **e** InN(1$\bar{1}$01) and **f** InN(11$\bar{2}$2) surfaces as functions of μ_{In} and μ_H. The origins of μ_H and μ_{In} correspond to H_2 molecules and bulk In (In droplet) at $T = 0$ K, respectively. Stable region of the surfaces with hydrogen is emphasized by shaded area. Open circles indicate μ_H and μ_{Ga} with $p_{H_2} = 76$ and $p_{In} = 5 \times 10^{-4}$ Torr, respectively, ranging from 770 to 900 K. Stable region of the surfaces with hydrogen is emphasized by shaded area. Top views of surface structures are also shown. Notations of circles are the same as those in Fig. 4.17. The positions of H atoms in the H-terminated surfaces are marked by crosses

The surfaces with NH and NH_2 are also stabilized under H-rich conditions on InN(11$\bar{2}$0) and (11$\bar{2}$2) surfaces as shown in Fig. 4.27d and f, respectively. If we assume the pressures of H_2 and In as $p_{H_2} = 76$ and $p_{In} = 5.0 \times 10^{-4}$ Torr for $770 - 900$ K, respectively, [open circles in Fig. 4.27d and f] the H-terminated surfaces with NH_2 on InN(11$\bar{2}$0) and InN(11$\bar{2}$2) surfaces (N–H + In–NH_2 and In–NH_2–In + N–H, respectively) are stabilized over the wide range of μ_H and μ_{In}. Since there are many excess electrons on these surfaces, the stability of InN surfaces on nonpolar and semipolar orientations is quite different from that on GaN surfaces. Due to low growth temperatures of InN, the surfaces with large number of N–H bonds become the most favorable configuration even though many excess electrons are generated by N–H bonds. The EC rule [41] is no longer satisfied on semipolar InN(11$\bar{2}$2) surfaces. This trend can be seen for InN(1$\bar{1}$00) and (1$\bar{1}$01) surfaces shown in Fig. 4.27c and e.

4.3.6 Surface Phase Diagrams for Hydrogen Adsorption on InN Surfaces

Figure 4.28 displays the calculated surface phase diagrams of H-adsorbed InN surfaces for polar (0001), polar (000$\bar{1}$), nonpolar (1$\bar{1}$00), nonpolar (11$\bar{2}$0), semipolar (1$\bar{1}$01) and semipolar (11$\bar{2}$2) orientations as functions of temperature and In BEP, assuming an H_2 pressure ($p_{H_2} = 76$ Torr) that corresponds to H-rich conditions [18, 23]. Similar to GaN surfaces, the adsorption of H atoms exhibits a different surface phase diagram compared with those without H atoms, as shown in Figs. 4.18, 4.20 and 4.22. These surface phase diagrams demonstrate that the H-terminated surfaces with NH and NH_2 are stabilized at temperatures above 675–900 K. Therefore, the H-terminated surfaces, such as the N_{ad}–H + In–NH_2 on the InN(0001) surface [Fig. 4.28a], the 4N–H on the InN(0001) surface [Fig. 4.28b], the N–H + In–NH_2 on the InN(1$\bar{1}$00) and (11$\bar{2}$0) surfaces [Fig. 4.28c and d], and the In–NH_2–In + N–H on the InN(1$\bar{1}$01) and (11$\bar{2}$2) surfaces [Fig. 4.28e and f] always emerge regardless of the growth conditions. Because there are excess electrons on these surfaces, the InN surface stability of nonpolar and semipolar orientations is quite different from that of GaN surfaces. Because of low InN growth temperatures, the surfaces with a large number of N–H bonds become the most

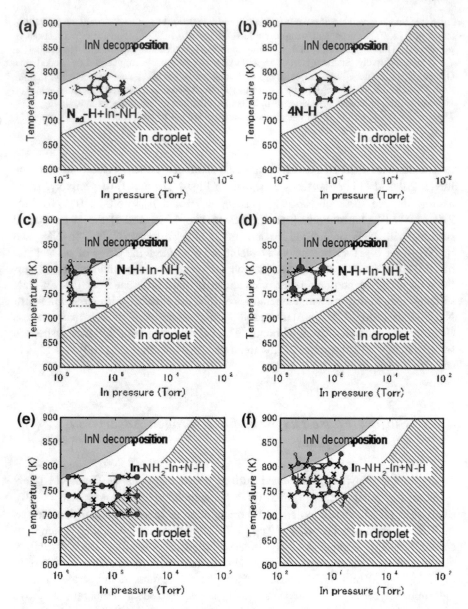

Fig. 4.28 Calculated surface phase diagram for H-adsorption on polar **a** InN(0001) and **b** InN($000\bar{1}$) surfaces, nonpolar **c** InN($1\bar{1}00$) and **d** InN($11\bar{2}0$) surfaces, and semipolar **e** InN($1\bar{1}01$) and **f** InN($11\bar{2}2$) surfaces as functions of temperature and In BEP under high H_2 pressure ($p_{H_2} = 76$ Torr) conditions. Top views of surface structures are also shown. Notations of circles are the same as those in Fig. 4.17. The positions of H atoms in the H-terminated surfaces are marked by crosses

favorable configurations even though many excess electrons are generated by N–H bonds. The EC rule [41] is no longer satisfied on semipolar surfaces. The absence of orientation dependence suggests that the growth kinetics on nonpolar and semipolar surfaces are similar to polar surfaces. Because the growth of InN on a InN(0001) surface is known to be prevented at high H_2 pressures, [65] the growth on nonpolar and semipolar surfaces is also inhibited due to the presence of hydrogen.

References

1. D.K. Biegelsen, R.D. Bringans, J.E. Northrup, L.-E. Swartz, Reconstructions of GaAs($\bar{1}\bar{1}\bar{1}$) surfaces observed by scanning tunneling microscopy. Phys. Rev. Lett. **65**, 452 (1990)
2. T. Ohno, Energetics of as dimers on GaAs(001) as-rich surfaces. Phys. Rev. Lett. **70**, 631 (1993)
3. J.E. Northrup, S. Froyen, Energetics of GaAs(100)–(2 × 4) and –(4 × 2) reconstructions. Phys. Rev. Lett. **71**, 22 (1993)
4. A. Kley, N. Moll, E. Pehlke, M. Scheffler, GaAs equilibrium crystal shape from first principles. Phys. Rev. B **54**, 8844 (1996)
5. J.E. Northrup, J. Neugebauer, R.M. Feenstra, A.R. Smith, Structure of GaN(0001): the laterally contracted Ga bilayer model. Phys. Rev. B **61**, 9932 (2000)
6. A. Ishii, First-principles study for molecular beam epitaxial growth of GaN(0001). Appl. Surf. Sci. **216**, 447 (2003)
7. Y. Kangawa, T. Ito, A. Taguchi, K. Shiraishi, T. Ohachi, A new theoretical approach to adsorption–desorption behavior of Ga on GaAs surfaces. Surf. Sci. **493**, 178 (2001)
8. T. Ito, H. Ishizaki, T. Akiyama, K. Nakamura, An ab initio-based approach to phase diagram calculations for GaAs(001) surfaces. e-J. Surf. Sci. Nanotech. **3**, 488 (2005)
9. H. Tatematsu, K. Sano, T. Akiyama, K. Nakamura, T. Ito, Ab initio based approach to initial growth processes on GaAs(111)B-(2 × 2) surfaces: self-surfactant effect of Ga adatoms revisited. Phys. Rev. B **77**, 233306 (2008)
10. T. Ito, N. Ishimure, T. Akiyama, K. Nakamura, Ab initio-based approach to adsorption–desorption behavior on the InAs(111)A heteroepitaxially grown on GaAs substrate. J. Cryst. Growth **318**, 72 (2011)
11. Y. Kangawa, Y. Matsuo, T. Akiyama, T. Ito, K. Shiraishi, K. Kakimoto, Theoretical approach to initial growth kinetics of GaN on GaN(001). J. Cryst. Growth **300**, 62 (2007)
12. T. Ito, T. Akiyama, K. Nakamura, Ab initio-based approach to structural change of compound semiconductor surfaces during MBE growth. J. Cryst. Growth **311**, 698 (2009)
13. T. Yamashita, T. Akiyama, K. Nakamura, T. Ito, Surface reconstructions on GaN and InN semipolar ($11\bar{2}2$) surfaces. Jpn. J. Appl. Phys. **48**, 120201 (2009)
14. T. Ito, T. Akiyama, K. Nakamura, An ab initio-based approach to the stability of GaN(0001) surfaces under Ga-rich conditions. J. Cryst. Growth **311**, 3093 (2009)
15. Y. Kangawa, T. Akiyama, T. Ito, K. Shiraishi, K. Kakimoto, Theoretical approach to structural stability of GaN: how to grow cubic GaN. J. Cryst. Growth **311**, 3106 (2009)
16. T. Akiyama, D. Ammi, K. Nakamura, T. Ito, Reconstructions of GaN and InN semipolar ($10\bar{1}1$) surfaces. Jpn. J. Appl. Phys. **48**, 100201 (2009)
17. T. Akiyama, D. Ammi, K. Nakamura, T. Ito, Surface reconstruction and magnesium incorporation on semipolar GaN(1101) surfaces. Phys. Rev. B **81**, 245317 (2010)
18. T. Akiyama, T. Yamashita, K. Nakamura, T. Ito, Stability of hydrogen on nonpolar and semipolar nitride surfaces: role of surface orientation. J. Cryst. Growth **318**, 79 (2011)
19. T. Ito, T. Akiyama, K. Nakamura, Ab initio-based approach to reconstruction, adsorption and incorporation on GaN surfaces. Semicond. Sci. Technol. **27**, 024010 (2012)

20. T. Akiyama, D. Obara, K. Nakamura, T. Ito, Reconstructions on AlN polar surfaces under hydrogen rich conditions. Jpn. J. Appl. Phys. **51**, 018001 (2012)
21. T. Akiyama, Y. Saito, K. Nakamura, T. Ito, Reconstructions on AlN nonpolar surfaces in the presence of hydrogen. Jpn. J. Appl. Phys. **51**, 048002 (2012)
22. T. Akiyama, K. Nakamura, T. Ito, Ab initio-based study for adatom kinetics on AlN(0001) surfaces during metal-organic vapor-phase epitaxy growth. Appl. Phys. Lett. **100**, 251601 (2012)
23. Y. Kangawa, T. Akiyama, T. Ito, K. Shiraishi, T. Nakayama, Surface stability and growth kinetics of compound semiconductors: an ab initio-based approach. Materials **6**, 3309 (2013)
24. Y. Takemoto, T. Akiyama, K. Nakamura, T. Ito, Systematic theoretical investigations on surface reconstruction and adatom kinetics on AlN semipolar surfaces. e-J. Surf. Sci. Nanotech. **13**, 239 (2015)
25. Y. Takemoto, T. Akiyama, K. Nakamura, T. Ito, Ab initio-based study for surface reconstructions and adsorption behavior on semipolar AlN(11$\bar{2}$2) surfaces during metal-organic vapor-phase epitaxy growth. Jpn. J. Appl. Phys. **54**, 0875502 (2015)
26. T. Akiyama, Y. Takemoto, K. Nakamura, T. Ito, Theoretical investigations of initial growth processes on semipolar AlN(11$\bar{2}$2) surfaces under metal–organic vapor-phase epitaxy growth condition. Jpn. J. Appl. Phys. **55**, 05FA06 (2016)
27. C.G. Van de Walle, J. Neugebauer, First-principles surface phase diagram for hydrogen on GaN surfaces. Phys. Rev. Lett. **88**, 066103 (2002)
28. H. Shu, X. Chen, R. Dong, X. Wang, W. Lu, Thermodynamic phase diagram for hydrogen on polar InP(111)B surfaces. J. Appl. Phys. **107**, 063516 (2010)
29. K. Yamada, N. Inoue, J. Osaka, K. Wada, *In situ* observation of molecular beam epitaxy of GaAs and AlGaAs under deficient As$_4$ flux by scanning reflection electron microscopy. Appl. Phys. Lett. **55**, 622 (1989)
30. T. Kojima, N.J. Kawai, T. Nakagawa, K. Ohta, T. Sakamoto, M. Kawashima, Layer-by-layer sublimation observed by reflection high-energy electron diffraction intensity oscillation in a molecular beam epitaxy system. Appl. Phys. Lett. **47**, 286 (1985)
31. E.M. Gibson, C.T. Foxon, J. Zhang, B.A. Joyce, Gallium desorption from GaAs and (Al, Ga) As during molecular beam epitaxy growth at high temperatures. Appl. Phys. Lett. **57**, 1203 (1990)
32. J.E. Northrup J. Neugebauer, Theory of GaN(10$\bar{1}$0) and (11$\bar{2}$0) surfaces. Phys. Rev. B **53**, R10477 (1996)
33. A.R. Smith, R.M. Feenstra, D.W. Greve, J. Neugebauer, J.E. Northrup, Reconstructions of the GaN(0001) surface. Phys. Rev. Lett. **79**, 3934 (1997)
34. J.E. Northrup, R. Di Felice, J. Neugebauer, Atomic structure and stability of AlN(0001) and (000$\bar{1}$) surfaces. Phys. Rev. B **55**, 13878 (1997)
35. J. Fritsch, O.F. Sankey, K.E. Schmidt, J.B. Page, Ab initio calculation of the stoichiometry and structure of the (0001) surfaces of GaN and AlN. Phys. Rev. B **57**, 15360 (1998)
36. C.D. Lee, Y. Dong, R.M. Feenstra, J.E. Northrup, J. Neugebauer, Reconstructions of the AlN (0001) surface. Phys. Rev. B **68**, 205317 (2003)
37. C.K. Gan, D.J. Srolovitz, First-principles study of wurtzite InN(0001) and (000$\bar{1}$) surfaces. Phys. Rev. B **74**, 115319 (2006)
38. D. Segev, C.G. van de Walle, Surface reconstructions on InN and GaN polar and nonpolar surfaces. Surf. Sci. **601**, L15 (2007)
39. H. Suzuki, R. Togashi, H. Murakami, Y. Kumagai, A. Koukitu, Theoretical analysis for surface reconstruction of AlN and InN in the presence of hydrogen. Jpn. J. Appl. Phys. **46**, 5112 (2007)
40. M.S. Miao, A. Janotti, C.G. van de Walle, Reconstructions and origin of surface states on AlN polar and nonpolar surfaces. Phys. Rev. B **80**, 155319 (2009)
41. M.D. Pashley, K.W. Haberern, W. Friday, J.M. Woodall, P.D. Kirchner, Structure of GaAs (001) (2 × 4)-c(2 × 8) determined by scanning tunneling microscopy. Phys. Rev. Lett. **60**, 2176 (1998)

42. F. Bernardini, V. Fiorentini, Macroscopic polarization and band offsets at nitride heterojunctions. Phys. Rev. B **57**, R9427 (1997)
43. P. Waltereit, O. Brandt, A. Trampert, H.T. Grahn, J. Menniger, M. Ramsteiner, M. Reiche, K. H. Ploog, Nitride semiconductors free of electrostatic fields for efficient white light-emitting diodes. Nature **406**, 865 (2000)
44. K. Nishizuka, M. Funato, Y. Kawakami, S. Fujita, Y. Narukawa, T. Mukai, Efficient radiative recombination from $11\bar{2}2$-oriented $In_xGa_{1-x}N$ multiple quantum wells fabricated by the regrowth technique, Appl. Phys. Lett. **85**, 3122 (2004)
45. K. Nishizuka, M. Funato, Y. Kawakami, Y. Narukawa, T. Mukai, Efficient rainbow color luminescence from $In_xGa_{1-x}N$ single quantum wells fabricated on $\{11\bar{2}2\}$ microfacets. Appl. Phys. Lett. **87**, 231901 (2005)
46. R. Sharma, P.M. Pattison, H. Masui, R. M. Farrel, T.J. Baker, B.A. Haskell, F. Wu, S. P. DenBaars, J.S. Speck, S. Nakamura, Demonstration of a semipolar $(10\bar{1}3)$ InGaN/GaN green light emitting diode, Appl. Phys. Lett. **87**, 231110 (2005)
47. T.J. Baker, B.A. Haskell, F. Wu, P.T. Fini, J.S. Speck, S. Nakamura, Characterization of planar semipolar gallium nitride films on spinel substrates. Jpn. J. Appl. Phys. **44**, L920 (2005)
48. A. Chakraborty, T.J. Baker, B.A. Haskell, F. Wu, J.S. Speck, S.P. DenBaars, S. Nakamura, U. K. Mishra, Milliwatt power blue InGaN/GaN light-emitting diodes on semipolar GaN templates. Jpn. J. Appl. Phys. **44**, L945 (2005)
49. M. Funato, T. Kotani, T. Kondou, Y. Kawakami, Y. Narukawa, T. Mukai, Tailored emission color synthesis using microfacet quantum wells consisting of nitride semiconductors without phosphors. Appl. Phys. Lett. **88**, 261920 (2006)
50. M. Ueda, K. Kojima, M. Funato, Y. Kawakami, Y. Narukawa, T. Mukai, Epitaxial growth and optical properties of semipolar $(11\bar{2}2)$ GaN and InGaN/GaN quantum wells on GaN bulk substrates. Appl. Phys. Lett. **89**, 211907 (2006)
51. J. Stellmach, M. Frentrup, F. Mehnke, M. Pristovsek, T. Wernicke, M. Kneissl, MOVPE growth of semipolar $(11\bar{2}2)$ AlN on m-plane $(10\bar{1}0)$ sapphire. J. Cryst. Growth **355**, 59 (2012)
52. Q.K. Xue, Q.Z. Xue, R.Z. Bakhtizin, Y. Hasegawa, I.S.T. Tsong, T. Sakurai, T. Ohno, Structures of GaN(0001)–(2 × 2), –(4 × 4), and –(5 × 5) surface reconstructions. Phys. Rev. Lett. **82**, 3074 (1999)
53. A.R. Smith, R.M. Feenstra, D.W. Greve, M.-S. Shin, M. Skowronski, J. Neugebauer, J.E. Northrup, GaN(0001) surface structures studied using scanning tunneling microscopy and first-principles total energy calculations. Surf. Sci. **423**, 70 (1999)
54. M.H. Xie, L.X. Zheng, X.Q. Dai, H.S. Wu, S.Y. Tong, A model for GaN ghost islands. Surf. Sci. **558**, 195 (2004)
55. V. Ramachandran, C.D. Lee, R.M. Feenstra, A.R. Smith, J.E. Northrup, D.W. Greve, Structure of clean and arsenic-covered GaN(0001) surfaces. J. Cryst. Growth **209**, 355 (2000)
56. R.M. Feenstra, J.E. Northrup, J. Neuegbauer, Review of structure of bare and adsorbate-covered GaN(0001) surfaces. MRS Internet J. Nitride Semicond. Res. **1**, 1234 (2002)
57. S. Vézian, F. Semond, J. Massies, D.W. Bullock, Z. Ding, P.M. Thibado, Origins of GaN (0001) surface reconstructions. Surf. Sci. **541**, 242 (2003)
58. L. Lahourcade, J. Renard, B. Gayral, E. Monroy, M.P. Chaivat, P. Ruterana, Ga kinetics in plasma-assisted molecular-beam epitaxy of GaN($11\bar{2}2$): effect on the structural and optical properties. J. Appl. Phys. **103**, 93514 (2008)
59. J.B. MacChesney, P.M. Bridenbaugh, P.B. O'Connor, Thermal stability of indium nitride at elevated temperatures and nitrogen pressures. Mater. Res. Bull. **5**, 783 (1970)
60. O. Ambacher, M.S. Brandt, R. Dimitrov, T. Metzger, M. Stutzmann, R.A. Fischer, A. Miehr, A. Bergmaier, G. Dollinger, Thermal stability and desorption of group III nitrides prepared by metal organic chemical vapor deposition. J. Vac. Sci. Technol. B **14**, 3532 (1996)

61. V.Y. Davydov, A.A. Klochikhin, R.P. Seisyan, V.V. Emptsev, S.V. Ivanov, F. Bechstedt, J. Furthmüller, H. Harima, A.V. Mudryi, J. Aderhold et al., Absorption and emission of hexagonal InN: evidence of narrow fundamental band gap. Phys. Status Solidi B **229**, R1 (2002)
62. J. Wu, W. Walukiewicz, K.M. Yu, J.W. Ager, E.E. Haller, H. Lu, W.J. Schaff, Y. Saito, Y. Nanishi, Unusual properties of the fundamental band gap of InN. Appl. Phys. Lett. **80**, 3967 (2002)
63. Y. Nanishi, Y. Saito, T. Yamaguchi, RF-molecular beam epitaxy growth and properties of InN and related alloys. Jpn. J. Appl. Phys. **42**, 2549 (2003)
64. Y. Saito, Y. Tanabe, T. Yamaguchi, N. Teraguchi, A. Suzuki, T. Araki, Y. Nanishi, Polarity of high-quality indium nitride grown by RF molecular beam epitaxy. Phys. Status Solidi B **228**, 13 (2001)
65. A. Koukitsu, T. Taki, N. Takahashi, H. Seki, Thermodynamic study on the role of hydrogen during the MOVPE growth of group III nitrides. J. Cryst. Growth **197**, 99 (1999)

Part II
Applications of Computational Approach to Epitaxial Growth of III-Nitride Compounds

Chapter 5
Thermodynamic Approach to InN Epitaxy

Yoshihiro Kangawa

In this chapter, influences of N/III ratio, growth orientation and total pressure on epitaxial growth processes of In(Ga)N are discussed. It is known that N/III ratio is essential parameter to grow In(Ga)N thin films [1, 2]. Figure 5.1 shows equilibrium vapor pressure as a function of reciprocal temperature [3, 4]. One can see the equilibrium vapor pressure of InN is higher than that of other III-nitride semiconductors. It implies that decomposition temperature of InN is lower than that of other III-nitride semiconductors. To perform high temperature growth of InN, Matsuoka et al. increased N/III ratio from typical 2,000 to 20,000 or more [1, 2]. The increase of NH_3 input partial pressure suppress the nitrogen decomposition during InN metal-organic vapor-phase epitaxy (MOVPE). That is, it is important to optimize N/III ratio especially for InN MOVPE. Recently, the other approaches to increase optimum growth temperature of In(Ga)N have been examined [5–10]. One approach is the growth of InGaN on nonpolar and semi-polar substrates. Figure 5.2 shows the normalized indium composition in InGaN relative to that of the (0001) plane as a function of the surface offcut angle. All the data are cited from Keller et al. [5] (orange squares), Yamada et al. [6] (blue triangles), Chichibu et al. [7] (pink inverted triangles), Wunderer et al. [8] (red circles), Wernicke et al. [9] (green diamonds), and Jönen et al. [10] (red diamond). Higher indium composition suggests that the growth orientations such as $(10\bar{1}1)$ and $(000\bar{1})$ are suitable for In(Ga)N high temperature growth since indium decomposition during growth seems suppressed at those growth surfaces. Furthermore, growth of InGaN on semi-polar and/or nonpolar surfaces are important to eliminate piezoelectric fields which affect the internal quantum efficiency of LEDs [11]. Another approach is InGaN growth

Y. Kangawa (✉)
Research Institutes for Applied Mechanics, Kyushu University, Fukuoka, Japan
e-mail: kangawa@riam.kyushu-u.ac.jp

© Springer International Publishing AG, part of Springer Nature 2018   95
T. Matsuoka and Y. Kangawa (eds.), *Epitaxial Growth of III-Nitride Compounds*,
Springer Series in Materials Science 269,
https://doi.org/10.1007/978-3-319-76641-6_5

Fig. 5.1 Equilibrium vapor
pressures of N_2 over AlN,
GaN and InN, the sum of As_2
and As_4 over GaAs, and sum
of P_2 and P_4 over InP [4].
Reproduced with permission
from Matsuoka [4]. Copyright
(1992) by Elsevier

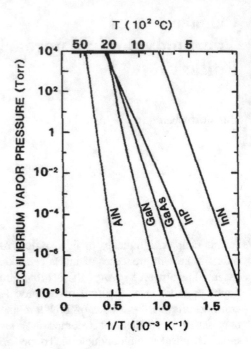

Fig. 5.2 Relationship
between the normalized In
composition relative to that of
(0001)+c plane and the
InGaN growth orientation
[5–10]

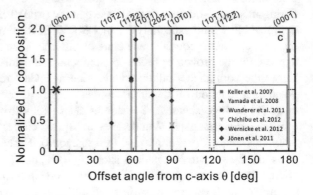

using pressurized-reactor MOVPE (PR-MOVPE) [12–14]. The PR-MOVPE is
effective for increasing optimum growth temperature of InGaN, since the decomposition of the material seems suppressed by the high total pressure. In this section,
influences of the N/III ratio, growth orientation and total pressure on the growth
phenomena are discussed from a viewpoint of thermodynamic approach. The calculation method for thermodynamic approach is described in Sect. 5.1. In Sect. 5.2,
stability of various InN surfaces is discussed. In Sect. 5.3, effect of total pressure on
the InN growth phenomena is described.

5.1 Thermodynamic Approach

Thermodynamic analysis is one of powerful analyzing methods to predict optimum growth conditions of compound semiconductors [15–20]. Figure 5.3 shows solid composition x of $In_xGa_{1-x}N$ alloy as a function of input In/(In+Ga) ratio, R_{In}, obtained by thermodynamic analysis [20]. Experimental data inserted in Fig. 5.3 were obtained by Matsuoka et al. [1] The calculation conditions were the same with those of experiments, i.e., input partial pressure of group-III gaseous sources, N/III ratio, and NH_3 decomposition ratio were 3×10^{-6} atm, 20,000 and 0.0 at 500 °C; 3.6×10^{-6} atm, 25,000 and 0.25 at 700 °C; and 1×10^{-5} atm, 5,000 and 0.35 at 800 °C, respectively. The total pressure was 0.1 atm, and carrier gas was N_2. One can see the calculation results agree well with those of experimental results. By the thermodynamic analysis, we can predict and discuss the influence of N/III ratio on the growth processes of group-III nitrides. However, the conventional thermodynamic analysis considers "gas−solid (bulk)" reaction instead of "gas−solid (surface)" reaction. In the following section, we explain how to incorporate the surface energy into the thermodynamic analysis.

5.1.1 Modeling InN MOVPE

In the typical InN MOVPE, trimethylindium (TMI) and NH_3 are used as source gases. The following thermal decomposition occurs near the substrate.

$$(CH_3)_3In(g) + \frac{3}{2}H_2(g) \rightarrow In(g) + 3CH_4(g), \qquad (5.1)$$

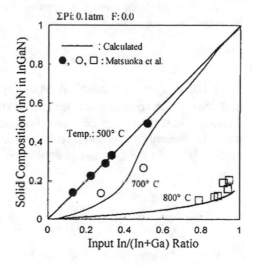

Fig. 5.3 Comparison between the theoretical results and the experimental results. The calculated conditions were the same with those of experiments: In input partial pressures, N/III ratio and α were 3×10^{-6} atm, 20000 and 0.0 at 500 °C; 3.6×10^{-6} atm, 25000 and 0.25 at 700 °C; and 1×10^{-5} atm, 5000 and 0.35 at 800 °C [17]. Reproduced with permission from Koukitu et al [17]. Copyright (1997) by Elsevier

$$NH_3(g) \rightarrow \frac{1}{2}\alpha N_2(g) + \frac{3}{2}\alpha H_2(g) + (1-\alpha)NH_3(g). \tag{5.2}$$

Here, α is the NH_3 decomposition ratio having the value of 0.25 [21]. The growth reaction at surface is written as

$$In(g) + NH_3(g) \rightarrow InN(s; \text{ surface}) + \frac{3}{2}H_2(g). \tag{5.3}$$

Here, $InN(s; \text{ surface})$ is the surface energy of InN. The second law of thermodynamics for the equilibrium condition is written as

$$\Delta G^0_{\text{surface--gas}} + RT \ln \left[\frac{a_{InN}(p_{H2})^{3/2}}{p_{In}p_{NH3}} \right] = 0, \tag{5.4}$$

where $\Delta G^0_{\text{surface--gas}}$ is the standard Gibbs energy of the reaction for (5.3); R is the gas constant; T is growth temperature; p is gas–surface equilibrium partial pressures; and a_{InN} is the activity of $InN(s; \text{ surface})$ that takes a value of 1 for pure substances. The assumption of a stoichiometric growth condition is written as

$$\Delta p_{In} = \Delta p_{NH3}, \tag{5.5}$$

$$\Delta p_{In} = -\frac{2}{3}\Delta p_{H2}, \tag{5.6}$$

where Δp represents the difference in the input partial pressure p^0 and the gas –surface equilibrium partial pressure p. By solving (5.4) under the stoichiometric growth conditions described in (5.5) and (5.6), the gas–surface equilibrium partial pressures p_{In}, p_{NH3}, and p_{H2} are obtained.

The standard Gibbs energy of reaction for (5.3), which is needed to solve (5.4), is written as

$$\Delta G^0_{\text{surface--gas}} = \Delta G^0_{\text{bulk--gas}} + \Delta E_{\text{surface--bulk}}. \tag{5.7}$$

Here, $\Delta G^0_{\text{bulk--gas}}$ is the standard Gibbs energy of reaction from the source gas to the bulk state, which corresponds to that used in the conventional analysis [15–17], and $\Delta E_{\text{surface--bulk}}$ is the energy difference between the bulk state and the surface state, which is obtained by ab initio calculations. In the following section, computational method to obtain $\Delta E_{\text{surface--bulk}}$ is described.

5.1.2 Surface Energy Calculation

In the ab initio calculation, a slab model composed of crystalline layers and vacuum region with two-dimensional periodic boundary conditions is used to investigate surface phenomena (See Fig. 5.4d). In the slab model, there are two surfaces, i.e., top and bottom surfaces. Dangling bonds in the bottom surface are generally passivated with fictitious hydrogen to form perfect covalent bonds [22]. In case of polar surfaces, surface energies of top and bottom surface are different. In order to investigate physical phenomena on top surface, we have to exclude the surface energy of bottom side from the total energy of the slab model. Figure 5.4 shows schematics for extracting surface energies [23, 24]. Figure 5.4a is a slab model of cubic GaN having nitrogen terminated (001) and $(00\bar{1})$ surfaces. Both side are passivated by fictitious hydrogen, and have a same geometry. Therefore, we can calculate the surface energy of one side by dividing by two. Figure 5.4b shows wedge-shaped model of cubic GaN surrounded by (111), $(\bar{1}\bar{1}1)$ and $(00\bar{1})$ N. The surface geometry of $(\bar{1}\bar{1}1)$ and (111) are the same, and the surface energy of $(00\bar{1})$ N is already known by the calculation of model (a). Therefore, we can extract the surface energy of (111). By using of model (c), the surface energy of $(\bar{1}\bar{1}1) \approx (000\bar{1})$, i.e., bottom surface of the slab model, can be obtained.

By using of the surface energy of the bottom side of the slab model, σ_{bottom}, the energy difference between the bulk state and the surface state is defined as

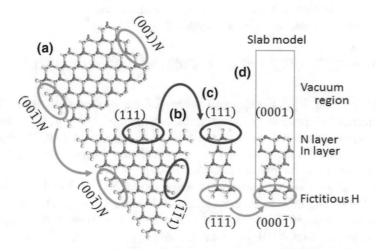

Fig. 5.4 Cross-sectional view of the wedge-shaped structures for extracting surface energies: **a** zinc-blende (ZB-) InN slab used to determine the energy of the passivated {001} surface, **b** ZB-InN triangular wedge used to determine the energy of the (111) surface, **c** ZB-InN slab used to determine the energy of the $(\bar{1}\bar{1}1)$ surface, and wurtzite (WZ-) InN slab used to determine the energy of the (0001) surface [23, 24]

$$\Delta E_{\text{surface-bulk}} = \left\{ \left[E_{\text{slab}} + n_N^{\text{ad}} \mu_{\text{In}}^{\text{InN(bulk)}} + n_{\text{In}}^{\text{ad}} \mu_N^{\text{InN(bulk)}} \right] \right.$$

$$- \left[\left(n_{\text{InN}}^{\text{slab}} + n_N^{\text{ad}} + n_{\text{In}}^{\text{ad}} \right) \mu_{\text{InN}}^{\text{InN(bulk)}} + n_H^{\text{ad}} \left(\mu_{\text{NH3}} - \mu_N^{\text{InN(bulk)}} \right) \right] / 3 \quad (5.8)$$

$$\left. + A_{\text{slab}} \sigma_{\text{bottom}} \right] \right\} N_A / \left(n_{\text{InN}}^{\text{top}} + n_N^{\text{ad}} + n_{\text{In}}^{\text{ad}} \right)$$

Here, E_{slab} is the total energy of the surface slab model; $\mu_{\text{In}}^{\text{InN(bulk)}}$, $\mu_N^{\text{InN(bulk)}}$, $\mu_{\text{InN}}^{\text{InN(bulk)}}$ are the chemical potentials of In, N, InN in InN(bulk), respectively; μ_{NH3} is the total energy of an ammonia molecule; $n_{\text{In}}^{\text{ad}}, n_N^{\text{ad}}, n_H^{\text{ad}}$ are the numbers of In, N, and H adatoms in the calculation cell, respectively; $n_{\text{InN}}^{\text{slab}}$ is the number of InN formula units in the calculation cell; $n_{\text{InN}}^{\text{top}}$ is the number of InN formula unit of the topmost layers; N_A is Avogadro's number; and A_{slab} is the surface area of slab model.

5.2 Surface Phase Diagram of InN Under MOVPE Condition

Surface reconstruction, or adsorption structure, depends on the growth conditions, such as the temperature and partial pressures. It is necessary to know which type of surface reconstruction appears on the growth surface to perform the thermodynamic analysis. The reconstructed surface with the lowest Gibbs energy should appear. Regarding a system consisting of an ideal surface and ambient gases as the energy-base, the Gibbs energies are written as

$$G = E_{\text{slab}}^{\text{recon}} - \left[E_{\text{slab}}^{\text{ideal}} + n_{\text{In}}^{\text{ad}} \mu_{\text{In}}^{\text{gas}} + n_N^{\text{ad}} \mu_{\text{NH3}}^{\text{gas}} + \left(\frac{1}{2} n_H^{\text{ad}} - \frac{3}{2} n_N^{\text{ad}} \right) \mu_{\text{H2}}^{\text{gas}} \right], \quad (5.9)$$

where $E_{\text{slab}}^{\text{recon}}$ and $E_{\text{slab}}^{\text{ideal}}$ are the total energy of the surface slab model for the reconstructed surfaces and an ideal surface, respectively, and $\mu_{\text{In}}^{\text{gas}}$, $\mu_{\text{NH3}}^{\text{gas}}$, and $\mu_{\text{H2}}^{\text{gas}}$ are the chemical potentials of In(g), NH$_3$(g), and H$_2$(g), respectively. The entropies of the gas molecules are calculated based on statistical mechanics as functions of the temperature and the partial pressures [25–27].

Figures 5.5a–d show the surface phase diagrams of InN (0001) +c, InN (000$\bar{1}$) −c, InN (10$\bar{1}$0) m, and InN (11$\bar{2}$0) a, respectively. A vertical axis is N/III ratio or NH$_3$ input partial pressure. A horizontal axis is temperature. Here, following growth parameters were considered, i.e., total pressure $\Sigma p_i = 1.0$, $p_{\text{In}}^0 = 1.0 \times 10^{-5}$ atm, $F = 0.0$, and $\alpha = 0.25$. The parameter F means the ratio of input hydrogen to nitrogen carrier gas. In the case of InN (0001) +c, and InN (10$\bar{1}$0) m, it was found that In terminated surfaces are stable under the typical growth conditions (see Fig. 5.5a, c). On the other hand, N−H terminated surfaces are stable at high temperatures though In-rich surfaces appears at low temperatures in the case of InN (000$\bar{1}$) −c, and InN

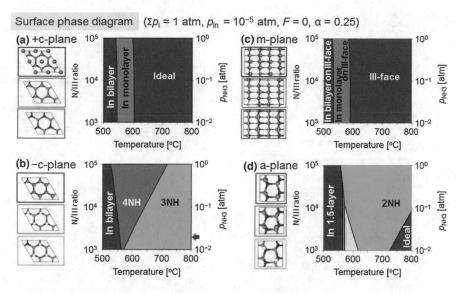

Fig. 5.5 Surface phase diagram of **a** InN (0001) +c, **b** InN $(000\bar{1})$ −c, **c** InN $(10\bar{1}0)$ m, and **d** InN $(11\bar{2}0)$ a planes. In input partial pressure p_{In} is constant (= 1×10^{-5} atm). The vertical axis is NH_3 input partial pressure p_{NH3} or N/III ratio. Blue arrow shows the condition of N/III = 2000

$(11\bar{2}0)$ a (see Fig. 5.5b, d). Figures 5.6a–d show the surface phase diagrams of InN (0001) +c, InN $(000\bar{1})$ −c, InN $(10\bar{1}0)$ m, and InN $(11\bar{2}0)$ a, respectively. In these figures, a vertical axis is In input partial pressure instead of NH_3 input partial pressure. Here, NH_3 input partial pressure is constant, i.e., p_{NH3}^0= 0.2 atm. Similar tendency is seen in these figures. That is, N−H terminated surfaces appears on InN $(000\bar{1})$ −c and InN $(11\bar{2}0)$a at high temperatures, though In terminated surfaces appears on InN (0001) +c and InN $(10\bar{1}0)$ m. However, definite phase transition temperatures are different from each other. Blue arrows in Fig. 5.5b show the condition of N/III = 2000, for example, the transition temperature from 4NH to 3NH surface phase is about 590 °C in former case while that is about 610 °C in latter case. This implies that it is important to analyze the experimental data by p_{In}^0 and p_{NH3}^0 individually, though the data are frequently arranged by N/III ratio. Furthermore, the phase transition temperature between the N–H terminated surfaces is more sensitive to NH_3 input partial pressure (see Fig. 5.5b) than In input partial pressure (see Fig. 5.6b). On the other hand, that between In terminated surfaces are more sensitive to p_{In}^0 and p_{NH3}^0. This is because NH_3 input partial pressure or its chemical potential in gas phase influences on the stability of N−H adsorbates, while stability of In adatoms mainly depends on In input partial pressure or its chemical potential in gas phase. In other words, the results suggest that the growth temperature range varies with In input partial pressure in case of InN (0001) +c and InN $(10\bar{1}0)$ m, while it depends on NH_3 input partial pressure in case of InN $(000\bar{1})$ −c and InN $(11\bar{2}0)$a.

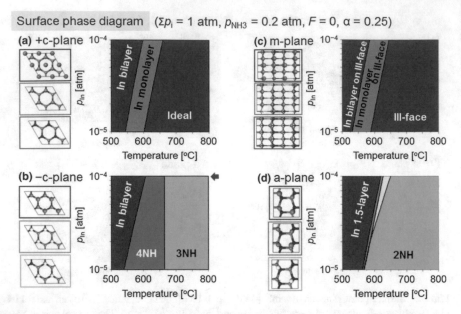

Fig. 5.6 Surface phase diagram of **a** InN (0001) +c, **b** InN $(000\bar{1})$ −c, **c** InN $(10\bar{1}0)$ m, and **d** InN $(11\bar{2}0)$ a planes. NH$_3$ input partial pressure p_{NH3} is constant (= 0.2 atm). The vertical axis is In input partial pressure p_{In}. Blue arrow shows the condition of N/III = 2000

Figures 5.7a–d show the gas−solid (surface) equilibrium partial pressure of In (blue and red solid line) as a function of temperature for InN (0001) + c, InN $(000\bar{1})$ −c, InN $(10\bar{1}0)$ m and InN $(11\bar{2}0)$ a, respectively. Due to the variable range of $\mu_{In}^{InN(bulk)}$ and $\mu_{N}^{InN(bulk)}$ in (5.8), the upper limit (blue solid line) and the lower limit (red solid line) appears in the figures. Black solid line and black dashed line show the gas−solid (bulk) equilibrium partial pressure of In and the input partial pressure of In, respectively. The growth conditions used for the analysis are as follows: $\Sigma p_i = 1.0$, $p_{In}^0 = 1.0 \times 10^{-5}$ atm, N/III = 20000 ($p_{NH3}^0 = 0.2$ atm), $F = 0.0$, and $\alpha = 0.25$. In the figures, there are discontinuous points in the p_{In} curves, i.e., blue solid lines and red solid lines. This is because there are changes in surface energies before and after the surface phase transitions. Here, the driving force for deposition of the material is defined as $\Delta p = p_{In}^0 - p_{In}$. If $\Delta p > 0$, epitaxial growth is able to proceed. If $\Delta p < 0$, however, epitaxial growth is difficult or would not proceed under the growth conditions. In case of InN (0001) +c, Δp changes from positive to negative value at 600 °C due to the transition from In monolayer surface to ideal surface (see Fig. 5.7a). This results suggest that the epitaxial growth proceed when In monolayer surface appears in case of InN (0001) +c, while it would not proceed when ideal surface appears. This is because ideal surface is unstable at high temperatures. On the other hand, 4NH and 3NH surfaces are stable at relatively high temperatures in case of InN $(000\bar{1})$ −c (see Fig. 5.7b). Therefore optimum growth temperature of InN $(000\bar{1})$ −c would be higher than that of InN (0001) +c. The driving

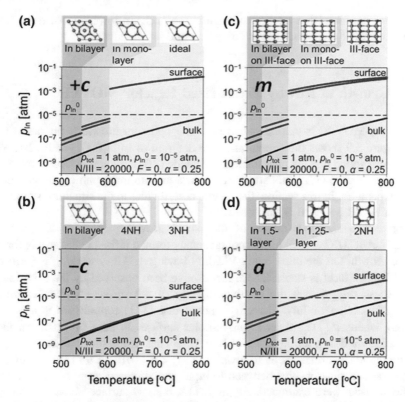

Fig. 5.7 Equilibrium partial pressure of indium p_{In} near the gas–solid (surface) interface. Blue and red solid line show the upper and lower limit of p_{In}. Black solid line shows the p_{In} near the gas –solid (bulk) interface. The calculation conditions are written in the figures

force for deposition Δp of each growth orientation is summarized in Fig. 5.8 [28]. The discontinuous points are due to the surface phase transitions. One can see the maximum growth temperature of InN $(000\bar{1})$ −c is the highest among the analyzed orientations. This calculation results agree well with the experimental results (see Fig. 5.2) [5–10, 13]. From this comparison, the feasibility of thermodynamic approach is confirmed.

Fig. 5.8 Driving force for deposition Δp $(= p_{In}^0 - p_{In})$ as a function of temperature. Discontinuous points appear due to the surface phase transitions

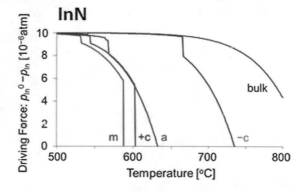

Moreover, it is found that the thermodynamic approach is a powerful tool to predict an optimum growth conditions of various compound semiconductors.

5.3 Growth of InN by Pressurized-Reactor MOVPE

InGaN PR-MOVPE is carried out to raise the optimum growth temperature [12–14]. Figure 5.9 shows the structural phase diagram of In N$(000\bar{1})$−c PR-MOVPE [13]. It was found that the metastable zinc blende (ZB) structure is incorporated into the stable wurtzite (WZ)-InN structure under a high In input partial pressure and a low temperature. In this section, the factors contributing to the structural stability of InN during PR-MOVPE is discussed.

Figure 5.10 shows a model of the island growth process of ZB-InN and WZ-InN. $ABCABC...$ stacking is spontaneously formed if the $\{\bar{1}\bar{1}\bar{1}\}$ facet is formed during growth. On the other hand, WZ-InN having $ABAB...$ stacking is formed if the $\{10\bar{1}0\}$ m facet is stable. These facets have been observed by the experiments [13]. These suggest that the relative stability of the surfaces influences on the structural stability of InN. Here, it is noted that $(\bar{1}\bar{1}\bar{1})$ top surface is a N-polar surface, whereas $\{\bar{1}\bar{1}\bar{1}\}$ facet is an In-polar surface. In case of $\{10\bar{1}0\}$m facet, occupation ratio in a top layer is In: N = 1: 1. That is, the order of In coverage θ among the ideal surfaces is as follows: θ $(\bar{1}\bar{1}\bar{1})$ < θ $(10\bar{1}0)$ < θ $(1\bar{1}\bar{1})$. From these considerations, relationships between In input partial pressure and relative stability of the surfaces were examined. Figure 5.11a, b show surface phase diagrams of WZ–InN $\{10\bar{1}0\}$ m and ZB–InN $\{\bar{1}\bar{1}\bar{1}\}$, respectively. The calculation conditions are as follows: $\Sigma p_i = 2400$ torr, flow rate of NH$_3$ is 15 slm, flow rate of inert gas is 5 slm, $F = 0.01$, and $\alpha = 0.25$. In Fig. 5.11, In bilayer or monolayer surfaces appears at low temperature region, while III face and ideal surfaces are observed at high temperature region. That is, In coverage becomes small with temperature increase.

Fig. 5.9 Phase diagram of InN PR-MOVPE. Hexagons and squares represent WZ and WZ/ZB mixed phase, respectively [13]

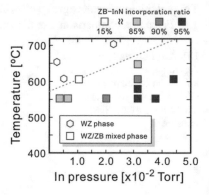

Fig. 5.10 Schematic of the model of island growth process. Orange and yellow circles denote In and N atoms, respectively

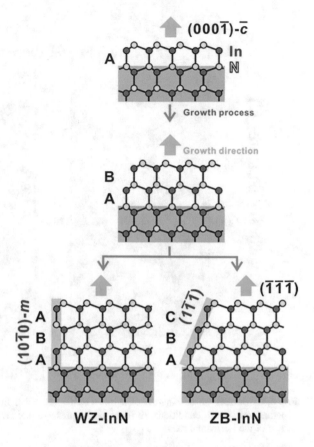

This is because the free energy of In atoms in the gas phase becomes small compared with the chemical potential of In adatoms on the surfaces.

Figure 5.12 shows the difference in surface energies between $\{10\bar{1}0\}$ m and $\{\bar{1}\bar{1}\bar{1}\}$ facet surfaces, i.e., $\Delta\sigma=\sigma\,(10\bar{1}0) - \sigma\,(1\bar{1}\bar{1})$. Here, σ represents the surface energy per surface area [23, 24]. Blue hexagon and red squares show the growth conditions of WZ–InN and ZB–InN, respectively [13]. In this growth regime, ZB–InN becomes more stable than WZ–InN under the conditions of low temperature and high In input partial pressure. Near the phase boundary between the WZ–InN and the ZB–InN mixture, it is found that the calculated $\Delta\sigma$ changes abruptly. Decrease in $|\Delta\sigma|$ causes the increase in probability of ZB–InN incorporation. The results imply that change in relative stability of $\{10\bar{1}0\}$ m and $\{1\bar{1}\bar{1}\}$ facet surfaces causes the incorporation of ZB phase during InN PR-MOVPE.

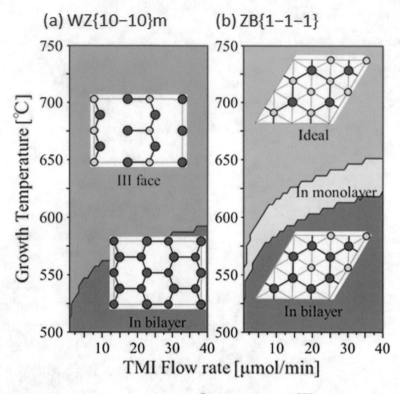

Fig. 5.11 Surface phase diagram of **a** InN $\{10\bar{1}0\}$ m, and **b** InN $\{\bar{1}\bar{1}\bar{1}\}$. To compare with the experimental growth conditions, In input partial pressure is converted into the gas flow rate and shown in the horizontal axis

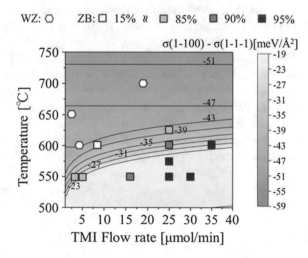

Fig. 5.12 Difference in surface energies: $\sigma\,(10\bar{1}0) - \sigma\,(1\bar{1}\bar{1})$. To compare with the experimental growth conditions, In input partial pressure is converted into the gas flow rate and shown in the horizontal axis. Hexagons and squares represent WZ and WZ/ZB mixed phase, respectively

References

1. T. Matsuoka, N. Yoshimoto, T. Sasaki, A. Katsui, Wide-gap semiconductor InGaN and InGaAlN grown by MOVPE. J. Electr. Mater. **21**, 157 (1992)
2. T. Matsuoka, H. Okamoto, M. Nakao, H. Harima, E. Kurimoto, Optical bandgap energy of wurtzite InN. Appl. Phys. Lett. **81**, 1246 (2002)
3. T. Matsuoka, H. Tanaka, T. Sasaki, A. Katsui, Wode-gap semiconductor (In, Ga)N. Inst. Phys. Conf. Set. **106**, 141 (1989)
4. T. Matsuoka, Current status of GaN and related compounds as wide-gap semiconductors. J. Cryst. Growth **124**, 433 (1992)
5. S. Keller, N.A. Fichtenbaum, M. Furukawa, J.S. Speck, S.P. DenBaars, U.K. Mishra, Growth and characterization of N-polar InGaN/GaN multiquantum wells. Appl. Phys. Lett. **90**, 191908 (2007)
6. H. Yamada, K. Iso, M. Saito, H. Hirayama, N. Fellows, H. Masui, K. Fujito, J.S. Speck, S. P. DenBaars, S. Nakamura, Comparison of InGaN/GaN light emitting diodes grown on m-plane and a-plane bulk GaN substrates. Phys. Status Solidi **2**, 89 (2008)
7. S.F. Chichibu, M. Kagaya, P. Corfdir, J.-D. Ganière, B. Deveaud-Pl èdran, N. Grandjean, S. Kubo, K. Fujito, Advantages and remaining issues of state-of-the-art m-plane freestanding GaN substrates grown by halide vapor phase epitaxy for m-plane InGaN epitaxial growth. Semicond. Sci. Technol. **27**, 024008 (2012)
8. T. Wunderer, M. Feneberg, F. Lipski, J. Wang, R.A.R. Leute, S. Schwaiger, K. Thonke, A. Chuvilin, U. Kaiser, S. Metzner, F. Bertram, J. Christen, G.J. Beirne, M. Jetter, P. Michler, L. Schade, C. Vierheilig, U.T. Schwarz, A.D. Dräger, A. Hangleiter, F. Scholz, Three-dimensional GaN for semipolarlight emitters. Phys. Status Solidi B **248**, 549 (2011)
9. T. Wernicke, L. Schade, C. Netzel, J. Rass, V. Hoffmann, S. Ploch, A. Knauer, M. Weyers, U. Schwarz, M. Kneissl, Indium incorporation and emission wavelength of polar, nonpolar and semipolar InGaN quantum wells. Semicond. Sci. Technol. **27**, 024014 (2012)
10. H. Jönen, U. Rossow, H. Bremers, L. Hoffmann, M. Brendel, A.D. Dräger, S. Metzner, F. Bertram, J. Christen, S. Schwaiger, F. Scholz, J. Thalmair, J. Zweck, A. Hangleiter, Indium incorporation in GaInN/GaN quantum well structures on polar and nonpolar surfaces. Phys. Status Solidi B **248**, 600 (2011)
11. T. Takeuchi, C. Wetzel, S. Yamaguchi, H. Sakai, H. Amano, I. Akasaki, Y. Kaneko, S. Nakagawa, Y. Yamaoka, N. Yamada, Determination of piezoelectric fields in strained GaInN quantum wells using the quantum-confined Stark effect. Appl. Phys. Lett. **73**, 1691 (1998)
12. T. Matsuoka, Y. Liu, T. Kimura, Y. Zhang, K. Prasertsuk, R. Katayama, Paving the way to high-quality indium nitride: the effects of pressurized reactor. Proc. SPIE **7945**, 794519 (2011)
13. T. Kimura, K. Prasertsuk, Y. Zhang, Y. Liu, T. Hanada, R. Katayama, T. Matsuoka, Phase diagram on phase purity of InN grown pressurized-reactor MOVPE. Phys. Status Solidi C **9**, 654 (2012)
14. T. Iwabuchi, Y. Liu, T. Kimura, Y. Zhang, K. Prasertsuk, H. Watanabe, N. Usami, R. Katayama, T. Matsuoka, Effect of phase purity on dislocation density of pressurized-reactor metalorganic vapor phase epitaxy grown InN. Jpn. J. Appl. Phys. **51**, 04DH02 (2012)
15. A. Koukitu, N. Takahashi, H. Seki, Thermodynamic study on metalorganic vapor-phase epitaxial growth of group III nitrides. Jpn. J. Appl. Phys. **36**, L1136 (1997)
16. A. Koukitu, H. Seki, Thermodynamic analysis on molecular beam epitaxy of GaN, InN and AlN. Jpn. J. Appl. Phys. **36**, L750 (1997)
17. A. Koukitu, S. Hama, T. Taki, H. Seki, Thermodynamic analysis of hydride vapor phase epitaxy of GaN. Jpn. J. Appl. Phys. **37**, 762 (1998)
18. K. Hanaoka, H. Murakami, Y. Kumagai, A. Koukitu, Thermodynamic analysis on HVPE growth of InGaN ternary alloy. J. Cryst. Growth **318**, 441 (2011)

19. Y. Kumagai, K. Takemoto, T. Hasegawa, A. Koukitu, H. Seki, Thermodynamics on tri-halide vapor-phase epitaxy of GaN and $In_xGa_{1-x}N$ using $GaCl_3$ and $InCl_3$. J. Cryst. Growth **231**, 57 (2001)
20. A. Koukitu, N. Takahashi, T. Taki, H. Seki, Thermodynamic analysis of the MOVPE growth of $In_xGa_{1-x}N$. J. Cryst. Growth **170**, 306 (1997)
21. T. Yayama, Y. Kangawa, K. Kakimoto, A. Koukitu, Theoretical analyses of In incorporation and compositional instability in coherently grown InGaN thin films. Phys. Status Solidi C **7**, 2249 (2010)
22. K. Shiraishi, A new model approach for electronic structure calculation of polar semiconductor surface. J. Phys. Soc. Jpn. **59**, 3455 (1990)
23. S.B. Zhang, Su-Huai Wei, Surface energy and the common dangling bond rule for semiconductors. Phys. Rev. Lett. **92**, 086102 (2004)
24. C.E. Dreyer, A. Janotti, C.G. Van de Walle, Absolute surface energies of polar and nonpolar planes of GaN. Phys. Rev. B **89**, 081305 (2014)
25. Y. Kangawa, T. Ito, A. Taguchi, K. Shiraishi, T. Ohachi, A new theoretical approach to adsorption–desorption behavior of Ga on GaAs surfaces. Surf. Sci. **493**, 178 (2001)
26. T. Ito, T. Akiyama, K. Nakamura, An ab initio-based approach to the stability of GaN(0001) surfaces under Ga-rich conditions. J. Cryst. Growth **311**, 3093 (2009)
27. Y. Kangawa, T. Akiyama, T. Ito, K. Shiraishi, T. Nakayama, Surface stability and growth kinetics of compound semiconductors: an ab initio-based approach. Materials **6**, 3309 (2013)
28. A. Kusaba, Y. Kangawa, P. Kempisty, K. Shiraishi, K. Kakimoto, A. Koukitu, Advances in modeling semiconductor epitaxy: contribution of growth orientation and surface reconstruction to InN MOVPE. Appl. Phys. Express **9**, 125601 (2016)

Chapter 6
Atomic Arrangement and In Composition in InGaN Quantum Wells

Yoshihiro Kangawa

In this section, atomic arrangement and indium incorporation in InGaN epitaxial layers are discussed. Chichibu et al. have studied why In-containing (Al, In, Ga)N films exhibit a defect-insensitive emission probability [1]. They concluded that localizing valence states associated with atomic condensates of In–N preferentially capture holes. Figure 6.1 shows a schematic of a trapped hole in InGaN films. A hole is trapped by an In–N−In–N−In–zigzag chain [2, 3] spontaneously formed along $[11\bar{2}0]$ direction in statistically homogeneous $In_{0.15}Ga_{0.85}N$ alloy. The holes form localized excitons to emit the light, though some of the excitons recombine at non-radiative centers. Therefore, it is important to discuss the atomic arrangement in InGaN epitaxial layers from a viewpoint of light-emitting-diode (LED) development. On the other hand, Suski et al. have investigated the relationship between the emission wavelength and structures of In(Ga)N/GaN short-period superlattices (SLs) [4]. Figure 6.2 shows the band gap energies and the photoluminescence (PL) energies of polar lIn(Ga)N/nGaN SLs [4–6]. Here, the n is a number of monolayers (MLs). That is, the specimen or calculation SL model is composed of alternately stacked l ML of In(Ga)N and n MLs of GaN. The dashed curve corresponds to the calculations performed for $In_xGa_{1-x}N$ quasi-random alloys [7, 8] with the asterisk marking the pure GaN [9]. In Fig. 6.2, for example, indium composition x of lInN/nGaN ($n = 3$) SLs is 0.25 (= 1/4) since the SLs composed of l ML of InN and 3 MLs of GaN. The calculated E_g of the SL model is less than 2.0 eV though that of $In_xGa_{1-x}N$ ($x = 0.25$) quasi-random alloy is about 2.5 eV. This implies that red shift of photoluminescence occurs by forming short-period SLs in comparison with that of bulk alloy. That is, short-period lIn (Ga)N/nGaN SLs enable band gap engineering in the blue-green range of the

Y. Kangawa (✉)
Research Institutes for Applied Mechanics, Kyushu University, Fukuoka, Japan
e-mail: kangawa@riam.kyushu-u.ac.jp

© Springer International Publishing AG, part of Springer Nature 2018
T. Matsuoka and Y. Kangawa (eds.), *Epitaxial Growth of III-Nitride Compounds*,
Springer Series in Materials Science 269,
https://doi.org/10.1007/978-3-319-76641-6_6

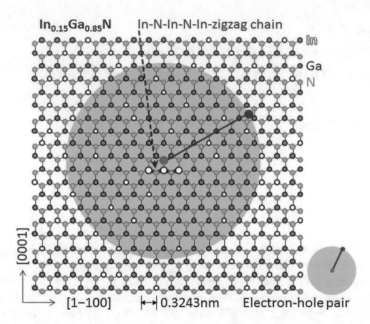

Fig. 6.1 A schematic representation of trapped hole in $In_{0.15}Ga_{0.85}N$ alloy. A hole trapped by an In–N–In–N–In-zigzag chain [2, 3] spontaneously formed along the $[11\bar{2}0]$ direction in statistically homogeneous $In_{0.15}Ga_{0.85}N$ alloy [1]

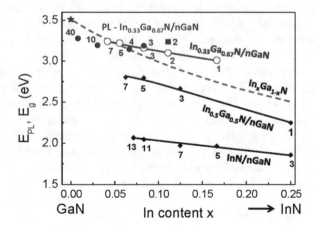

Fig. 6.2 Calculated band gap energies E_g of $In_{0.33}Ga_{0.67}/nGaN$ SLs (blue empty circles) in comparison with that for $In_{0.5}Ga_{0.5}N/nGaN$ and $InN/nGaN$ SLs (black diamonds) and with the PL energies E_{PL} (red closed circles and squares) [4]. The dashed curve corresponds to the calculations performed for $In_xGa_{1-x}N$ quasi-random alloys [7, 8] with the asterisk marking the pure GaN [9]. Reproduced with permission from Suski et al. [4]. Copyright (2014) by American Institute Physics

spectrum. It is also an important phenomenon to apply to the LED development. In Sect. 6.1, atomic arrangements in InGaN layer is discussed. In Sect. 6.2, indium incorporation in InGaN/nGaN SLs is reviewed.

Fig. 6.3 Cohesive energy as
a function of the atomic
number in the calculation
model (black squares
corresponds to the number of
atoms = 64, 512, 1728, and
4096) [11]

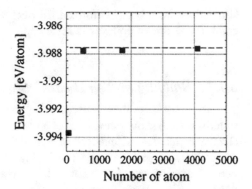

An atomic number of the model for ab initio calculation is limited. To analyze the atomic arrangement and indium incorporation in InGaN alloy, it is necessary to use a calculation model with a large atomic number. Saito and Arakawa investigated the phase stability of $In_xGa_{1-x}N$ based on regular-solution model using x-dependent interaction parameter Ω $(= -2.11x + 7.41$ kcal/mol) estimated from results of valence-force-field calculation and obtained a slightly deviated miscibility gap from a symmetric one [10]. However, these studies focus on the stability in bulk state and do not address those in the thin-film state. To study thermodynamic stability in thin-film state, especially the lattice constraint from the bottom layer must be incorporated. Kangawa et al. investigated the influence of lattice constraint on the excess energy curves of $In_xGa_{1-x}N$ on GaN and InN [11]. The empirical interatomic potential used in this work has been proposed by Ito et al. [12–14] (See Sects. 2.2 and 3.3 for more details) In the following section, the empirical interatomic potential was used to calculate the cohesive energies of the systems. A 4096-atom model was employed in the following works. The convergence of the system energy was confirmed by the energy calculation for $In_{0.5}Ga_{0.5}N$ as a function of the number of atoms in the system as shown in Fig. 6.3. The results suggest that the system energy becomes almost constant when the number of atoms in the system is larger than 512. That is, a 4096-atom model is feasible to investigate the stability of InGaN thin films.

6.1 Atomic Arrangement in InGaN

It is known that InGaN blue LEDs emit brilliant light although the threading dislocation density generated due to lattice mismatch is six orders of magnitude higher than that in conventional LEDs. As described above, Chichibu et al. reported that formation of In–N–In–N–In-zigzag chain enhance the generation of localized excitons to emit the light. In Sect 6.1.1, stability of tetrahedral clusters in InGaN epi-layer is discussed. Using the calculated energies of clusters as parameters, MC

simulations have been performed. In Sect. 6.1.2, density of In–N–In–N–In-zigzag
chain in the grown epilayer is reviewed.

6.1.1 Stability of Tetrahedral Clusters

Figures 6.4a ,b show schematics of a part of calculation model and cohesive energy
change in tetrahedral clusters as a function of surrounded indium composition,
respectively. A green tetrahedron in Fig. 6.4a denotes a cluster of attention. The
vertical axis of Fig. 6.4b shows the change in cohesive energy of this cluster.
A kind of clusters are illustrated in the right hand side of Fig. 6.4b. Within a blue
sphere drawn in Fig. 6.4a, there are 25 group-III atomic sites. The horizontal axis of
Fig. 6.4b shows indium composition in this sphere. Here, the total indium composition of a 4096-atomic model was kept to $x = 0.25$. That is, if indium composition in the blue sphere surrounding the green tetrahedron becomes large, that of
remaining area becomes small to keep composition constant. Here, the radius R of
the blue sphere is determined to 5 Å by considering the relationship between the
cohesive energy of clusters and the radius R. Figure 6.5a, b show the cohesive
energy of In_4N and Ga_4N clusters as a function of R, respectively. In these calculations, indium occupied all group-III sites in the sphere. One can see the
R would be more than 5 Å to incorporate the influence of local composition on the
stability of each cluster. In Fig. 6.4b, red, orange and green circles show the stability of In_4N, In_3Ga_1N and In_2Ga_2N cluster, respectively. They become unstable
when indium composition in the surrounding area becomes large. This implies
these clusters are unstable in the In-rich region. On the other hand, Ga-rich clusters
such as In_1Ga_3N and Ga_4N are stable regardless of the change in indium

Fig. 6.4 a Schematic of
InGaN cross-section. White
and orange circles show
group-III atoms and nitrogen,
respectively. Green triangle
shows the tetrahedral cluster
of attention. Blue circle
(sphere) shows the local area
surrounding the tetrahedral
cluster. **b** Change in cohesive
energy of the tetrahedral
cluster as a function of indium
composition in the local area.
Here, the composition of the
$In_xGa_{1-x}N$ calculation model
with 4096 atoms is 0.25

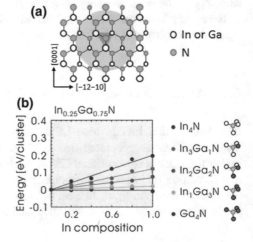

Fig. 6.5 Cohesive energy of the **a** In$_4$N and **b** Ga$_4$N clusters as a function of the radius of spherical local area (See Fig. 6.4a). In the calculation, indium occupies all group-III sites in the spherical local area

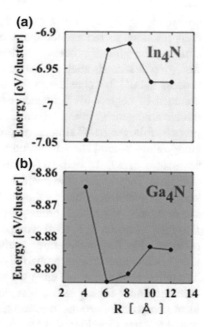

composition in the surrounding area. The approximate lines of the data of absolute cohesive energies are written as follows.

$$E_{coh} = A + Bx. \tag{6.1}$$

Here, x is the indium composition within a blue sphere in Fig. 6.4a. The constants A and B are written in Table 6.1. Using these data as MC parameters, growth simulations of InGaN epilayer were carried out.

6.1.2   Monte Carlo Simulation of InGaN MOVPE

Homo-epitaxial growth of InGaN (0001) with a wurtzite structure was considered. It is assumed that epitaxial growth proceeds principally by the following steps: (Step 1) In and Ga atoms adsorb one by one onto the (0001) surface. The surface is

Table 6.1 Cohesive energy of each cluster E_{coh} [eV/ cluster] is given as $A + Bx$. (See 6.1)

Cluster	A	B
In$_4$N	−7.2246	0.20097
In$_3$Ga$_1$N	−7.6762	0.12368
In$_2$Ga$_2$N	−8.1689	0.069737
In$_1$Ga$_3$N	−8.5626	0.018988
Ga$_4$N	−8.8766	−0.0008074

covered by group-III atoms, (Step 2) In and Ga atoms in the top layer exchange their positions to reduce configurational energy and (Step 3) N atoms cover the surface. The growth process is schematically drawn in Fig. 6.6. If growth rate is fast, growth kinetics such as atomic migration on the terrace and step-flow growth process should be considered. In this research, slow growth rate condition is assumed to neglect the influence of kinetic or nonequilibrium processes on the atomic arrangement in the grown layer. To reproduce the above mentioned growth process, following MC simulation procedure was proposed. (Step 1) Local indium composition around the adsorption site was computed. Here, the rectangular area with sides of 12, 12 and 10.6 Å in the length in the $[11\bar{2}0]$, $[1\bar{1}00]$ and $[0001]$ directions, respectively, is considered as a local area. The size of considered area seems sufficient considering the calculation results shown in Fig. 6.5. Then, the site-correlated adsorption probability of In and Ga was estimated using the relationships between the local composition and the driving force of deposition obtained by thermodynamic analyses (See Fig. 6.7) [15, 16]. Using the site-correlated adsorption probability, In or Ga was adsorbed on the site. (Step 2) The difference in cohesive energy before and after the site exchange, i.e., change in configurational energy ΔE_{ex}, was computed using the relationships between the local composition around the exchanging sites and cohesive energy of the tetrahedral clusters (See Figs. 6.4 and 6.8). Using the calculated ΔE_{ex}, site exchanging probability W_{ex} was computed by

$$W_{ex} = \frac{\exp(-\Delta E_{ex}/k_BT)}{1 + \exp(-\Delta E_{ex}/k_BT)}. \tag{6.2}$$

Here, k_B is the Boltzmann's constant, and T is the temperature. (Step 3) After the site-exchanging process, N atoms covered the surface, and then atomic configuration of group-III atoms was fixed. In the MC simulation, $150 \times 150 \times 44$ group-III atoms were deposited, and atomic configuration was analyzed.

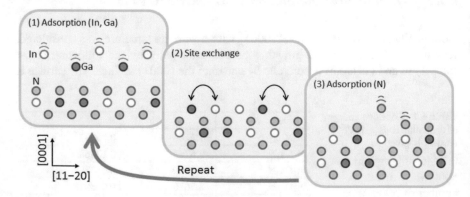

Fig. 6.6 Schematic of elementary process of InGaN growth

Fig. 6.7 a Schematic of local area of In or Ga adsorption site. **b** Driving force for deposition of InN ralatice to that of GaN as a function of indium composition in the local area. Δp_i is the driving force for deposition of element $-i$. Input partial pressure of group-III atoms $p_{III}^0 = 1 \times 10^{-5}$, that of ammonia $p_{NH3}^0 = 0.2$ atm, R_{In} $(= p_{In}^0/(p_{In}^0 + p_{Ga}^0)) = 0.25$, NH_3 decomposition ratio $\alpha = 0.25$, H_2 ratio in the carrier gas $F = 0.01$ and $\Sigma p_i = 1.0$ atm

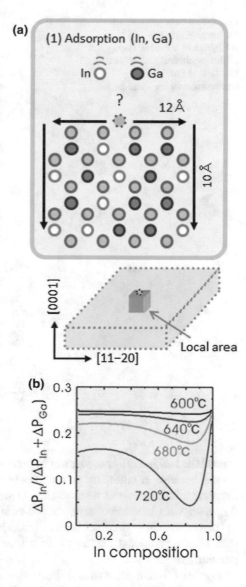

Figure 6.9a shows the schematic of atomic column in the [0001] direction. Figure 6.9b shows the relationship between the indium composition in the columns and their occupation ratio among the all columns in the grown models. Here, 5000 Monte Carlo steps (MCS) were carried out during the simulation. 1 MCS corresponds to 1 time of the site-exchanging event per one atom on the top surface. In the MC simulation, there were 150 × 150 atoms on the growth surface. Therefore, 5000 MCS means 5000 × 150 × 150 trials during 1 mono-layer (ML) growth. The number of 5000 MCS was determined to reproduce an equilibrium growth process (See Fig. 6.9c). In Fig. 6.9b, one can see the peak shift of the dispersion curve

Fig. 6.8 **a** Schematic of site
exchanging process.
b Change in cohesive energy
of the tetrahedral cluster as a
function of indium
composition in the local area

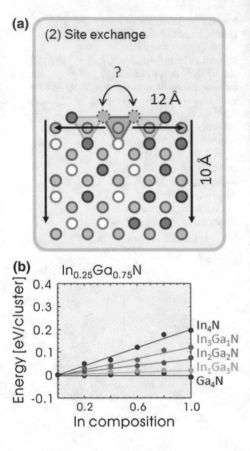

toward the low indium composition as increase of the growth temperature. The shift
occurs because equilibrium partial pressure of indium near the growth surface
decrease with increase of temperature. That is, driving force for deposition of InN
decreases with increase of temperature compared with that of GaN. Furthermore,
the dispersion curve becomes broad as the temperature decrease. This result implies
that the compositional fluctuation in a thin film is emphasized under a low temperature growth case.

   Figures 6.10a, c show an atomic arrangement in (0001) plane below 5 ML from
the top surface and schematic of the simulation model, respectively. The MC
simulation was performed under the conditions of 5000 MCS and 720 °C. For
comparison, atomic arrangement in the model simulated under the conditions of 0
MCS and 720 °C is shown in Fig. 6.10b. In Fig. 6.10b, indium atoms distribute
randomly since there was no site-exchanging trial during the MC simulation.
Orange, green and blue circles in Fig. 6.10a, b show the In–N–In–N–In-zigzag
chain along $\langle 11\bar{2}0 \rangle$, indium aggregated region and indium dispersal region,
respectively. It is noted that indium aggregation region appears in Fig. 6.10b while
indium dispersal region appears in Fig. 6.10a. This implies that indium atoms prefer

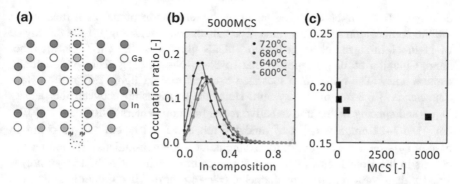

Fig. 6.9 **a** Schematic of atomic column in the [0001] direction, i.e., area surrounded by the dotted lines. **b** Relationship between indium composition in the column and their occupation ratio among all columns. **c** Relationship between MC step and the occupation ratio of the column having a maximum composition

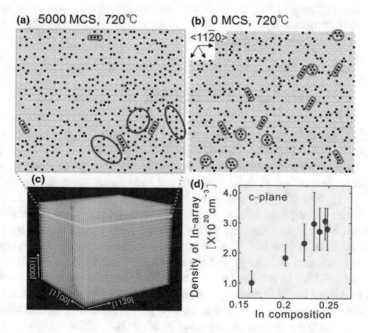

Fig. 6.10 Atomic arrangement in (0001) plane below 5 ML from the top surface in the model calculated under the conditions of **a** 5000 MCS, 720 °C and **b** 0 MCS, 720 °C. Black and white circles show In and Ga, respectively. Nitrogen atoms are not shown. **c** Schematic of MC simulation model. **d** Density of In–N–In–N–In-zigzag chain as a function of indium composition

to separate from each other under the equilibrium state since In-rich cluster is unstable as shown in Fig. 6.4. Moreover, In–N–In–N–In-zigzag chain certainly exist even under the equilibrium state as seen in Fig. 6.10a. Figure 6.10d shows density of In–N–In–N–In-zigzag chain in the simulation model as a function of indium composition in InGaN films. The calculation results suggest the density of In–N–In–N–In-zigzag chain in In-poor InGaN films is $\sim 10^{20}$ cm^{-3}. On the other hand, Chichibu et al. [1] reported that In–N–In–N–In-zigzag chain preferentially capture holes. The holes form localized excitons to emit light. The exciton Bohr diameter is ~ 6.8 nm in this system. Here, the relationship between the density D_{NRC} and spacing of the non-radiative recombination centers (NRC) are 1×10^{18} cm^{-3} and 7–11 nm; 5×10^{18} cm^{-3} and 4–6 nm, and 1×10^{19} cm^{-3} and 3–5 nm. In case of $D_{NRC} = 5 \times 10^{18}$ cm^{-3}, for example, there are non-radiative recombination centers in the Bohr radius of the exciton since the spacing of NRC (= 4–6 nm) is smaller than the exciton Bohr diameter (= ~ 6.8 nm). This situation causes the deterioration of the excitonic emission. If $D_{NRC} < 5 \times 10^{18}$ cm^{-3}, the spacing of NRC becomes more than 7 nm. In this situation, the In–N–In–N–In-zigzag chains can capture the excitons and emit light. The predicted density of In–N–In–N–In-zigzag chain ($\sim 10^{20}$ cm^{-3}) by the MC simulations is larger than the estimated D_{NRC} ($< 5 \times 10^{18}$ cm^{-3}) in InGaN films. The MC simulation results support the experimental consideration, i.e., InGaN can emit light if the spacing of NRC is larger than the exciton Bohr diameter.

6.2   In Incorporation in InGaN QWs

InGaN/GaN quantum structures are used in LEDs. Although the emission wavelength can be controlled from ultraviolet to violet, blue, green, and amber regions by increasing the In composition in the In$_x$Ga$_{1-x}$N active layer, device performance rapidly degrades at wavelengths beyond the blue-green spectral range. This is because it is difficult to grow a high crystalline quality In-rich InGaN by metal-organic vapor-phase epitaxy (MOVPE) or molecular beam epitaxy (MBE). Suski et al. [4] reported that red shift of photoluminescence occurs by forming short-period lIn(Ga)N/nGaN SLs in comparison with that of bulk alloy. This implies that emission of blue-green light is possible from In-poor InGaN if we use the short-period SLs. In this system, we can reduce the defect density, however it is difficult to control the indium composition in each layer. For example, it is reported that the lIn$_x$Ga$_{1-x}$N/nGaN SLs ($x = 0.33$) was grown instead of the intended $x = 1$ [4]. In this section, indium incorporation in InGaN hetero-epitaxial layer is reviewed. That is, influence of lattice constraint from bottom layer on the indium incorporation in InGaN films are discussed.

6.2.1 Effective Enthalpy of Mixing of Coherently Grown InGaN Layers

Figure 6.11 shows a schematic of calculation model composed of $16 \times 16 \times 16$ (= 4096) group-III atoms. Same number of nitrogen (4096) atoms are also incorporated in the model to form wurtzite structure. In the cohesive energy calculation using the empirical interatomic potentials (See Sect. 2.2), the following coherent growth conditions were considered: (I) the lattice parameter a of basal plane is fixed to be that of $In_yGa_{1-y}N$ bottom layer, (II) the lattice parameter c and displacement of atoms are varied to minimize the system energy. The convergence of the calculation model was confirmed by the energy calculation of $In_{0.5}Ga_{0.5}N$ as a function of the number of atoms in the model (See Fig. 6.3). The effective enthalpy of mixing $\Delta E_{eff}(x)$ is computed by

$$\Delta E_{eff}(x) = E(x) - \left\{ x E_{\text{bulk-InN}} + (1-x) E_{\text{bulk-GaN}} \right\}, \qquad (6.3)$$

where $E(x)$, $E_{\text{bulk-InN}}$ and $E_{\text{bulk-GaN}}$ are the cohesive energies of $In_xGa_{1-x}N$, bulk InN and bulk GaN, respectively. Figure 6.12 shows the effective enthalpy of mixing $\Delta E_{eff}(x)$ of $In_xGa_{1-x}N/In_yGa_{1-y}N$ (See Fig. 6.11). In case of bulk InGaN, the $\Delta E_{eff}(x)$ of $x = 1.0$ or 0.0 is zero since $E(x)$ is equal to $E_{\text{bulk-InN}}$ or $E_{\text{bulk-GaN}}$. In case of $In_xGa_{1-x}N/In_yGa_{1-y}N$, however, $\Delta E_{eff}(x)$ of $x = 1.0$ or 0.0 is not zero since $E(x)$ is not equal to $E_{\text{bulk-InN}}$ or $E_{\text{bulk-GaN}}$. The $\Delta E_{eff}(x)$ of $x = 0.0$ or 1.0 corresponds to the difference in cohesive energy between the coherently grown InN or GaN epilayers on $In_yGa_{1-y}N$ and that in bulk state. That is, influence of compressive or tensile stress on the cohesive energy of the system is incorporated in the effective enthalpy of mixing. The approximation curves of $\Delta E_{eff}(x)$ is written by the following polynomial.

$$\Delta E_{eff}(x) = a_0 + a_1 x + a_2 x^2 + a_3 x^3 + a_4 x^4 + a_5 x^5 + a_6 x^6. \qquad (6.4)$$

Fig. 6.11 Schematic of empirical interatomic potential calculation model

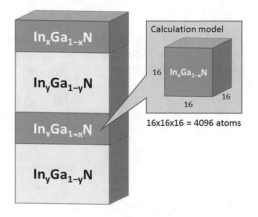

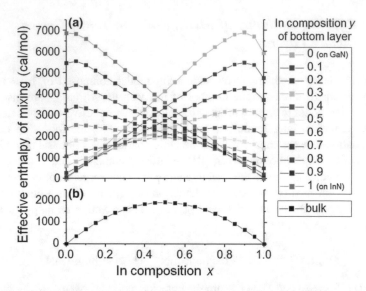

Fig. 6.12 Effective enthalpy of mixing curves of **a** $In_xGa_{1-x}N/In_yGa_{1-y}N$ SLs and **b** bulk InGaN

The constants a_i are written in Table 6.2. Saito and Arakawa investigated the phase stability of $In_xGa_{1-x}N$ based on regular-solution model using x-dependent interaction parameter Ω ($=-2.11x + 7.41$ kcal/mol) estimated from results of valence-force-field calculation and obtained a slightly deviated miscibility gap from a symmetric one [10]. When the enthalpy of mixing curve has a parabolic shape like Fig. 6.12b, the activities of binary compounds in the alloy a_{InN} and a_{GaN} are written as [17]

$$a_{InN} = x \exp\left[\frac{(1 - x^2)\Omega}{RT}\right], \tag{6.5}$$

$$a_{GaN} = (1 - x) \exp\left[\frac{x^2\Omega}{RT}\right], \tag{6.6}$$

using Ω, where R is the gas constant. However, the effective enthalpy of mixing curves shown in Fig. 6.12a are not a parabolic shape. In this case the a_{InN} and a_{GaN} are written as follows.

$$a_{InN} = x \exp\left[\frac{(1 - x)\frac{d(\Delta E_{eff}(x))}{dx} + \Delta E_{eff}(x)}{RT}\right], \tag{6.7}$$

$$a_{GaN} = (1 - x) \exp\left[\frac{-x\frac{d(\Delta E_{eff}(x))}{dx} + \Delta E_{eff}(x)}{RT}\right]. \tag{6.8}$$

Table 6.2 Constants of the approximation curves of the effective enthalpy of mixing $\Delta E_{\text{eff}} = a_0 + a_1 x + a_2 x^2 + a_3 x^3 + a_4 x^4 + a_5 x^5 + a_6 x^6$ (cal/mol)

y	a_0	a_1	a_2	a_3	a_4	a_5	a_6
0	0	8595.51696	-16070.44615	94601.2849	-223477.86767	240838.72742	-98549.1173
0.1	56.85162	7397.08096	-18477.33249	99299.88	-225005.39823	233221.5651	-91733.54018
0.2	258.36571	5793.45098	-14381.3333	73006.14981	-162070.24699	165931.38341	-64821.14642
0.3	587.02712	4910.14243	-17689.02469	77139.84448	-154941.02607	146243.59863	-53437.15841
0.4	1042.11147	4051.9364	-17969.12873	64196.29352	-113920.05989	98988.28402	-34349.71822
0.5	1632.43439	3874.40874	-25875.9249	83509.77399	-134998.46475	107590.38317	-34337.08802
0.6	2357.01481	3941.24541	-34889.85751	105902.32372	-162546.79615	123742.08475	-37631.31033
0.7	3224.23167	4387.87129	-47940.26774	145336.02076	-223611.37874	170587.42767	-51503.80333
0.8	4258.8793	4403.47828	-58073.22961	175836.03146	-271022.9371	207213.43485	-62406.64626
0.9	5465.71647	3200.03547	-56826.4819	163768.41037	-242837.27002	180185.46203	-52907.72376
1	6808.77056	4943.20836	-78807.05348	228729.3918	-342023.21648	255899.83837	-75561.50794

By using of these equations, contribution of the lattice constraint from the bottom layer, i.e., planar compressive or tensile stress in epilayer, is incorporated in the thermodynamic analysis [17].

6.2.2 Thermodynamic Analysis of InGaN Hetero-Epitaxy

Thermodynamic analyses of InGaN MOVPE under the conditions of input partial pressure of group-III gaseous sources $p_{III}^0 = 1 \times 10^{-5}$ atm; NH_3 decomposition rate is $\alpha = 0.25$; H_2 ratio in a carrier gas $F = 0.01$; and total pressure $\Sigma p = 1.0$ atm were carried out. Figure 6.13 shows the indium composition as a function of input indium source gas ratio R_{In} ($= p_{In}^0/(p_{In}^0 + p_{Ga}^0)$). The black and red squares show the results of bulk $In_xGa_{1-x}N$ case and $In_xGa_{1-x}N$ on $In_yGa_{1-y}N$ ($y = 0.3$) case. Here, $In_xGa_{1-x}N$ ($x < y = 0.3$) receive planar tensile stress while $In_xGa_{1-x}N$ ($x > y = 0.3$) receive planar compressive stress. Bond length of InN is longer than that of GaN. Therefore, In–N is stable (unstable) when the system receives planar tensile (compressive) stress. This phenomenon causes the composition pulling effect [18]. The composition pulling effect is seen in Fig. 6.13, i.e., indium composition of $In_xGa_{1-x}N$ films ($x < y = 0.3$) is larger than that in bulk state due to the tensile stress while indium composition of $In_xGa_{1-x}N$ films ($x > y = 0.3$) is smaller than that in bulk state due to the compressive stress. Figure 6.14 shows the phase diagram of $In_xGa_{1-x}N/In_yGa_{1-y}N$ MOVPE growth under the temperatures of 700, 740, and 780 °C. The blue, green and yellow regions indicate growth, unstable, and etching conditions, respectively. Here, "unstable" means the possibility of compositional fluctuation during growth. The calculation results suggest that the lattice constraint from bottom layer reduces the preferable growth regions (see Fig. 6.14a, b). This is because the planar compressive stress is accumulated in InGaN layer on GaN substrate. When the indium composition in bottom layer y increases, accumulated compressive stress in InGaN epilayer would be partially released. In this situation, InN becomes stable and preferable growth region would be increase (see Fig. 6.14b–d). Furthermore, the calculation results predicted that the InN epitaxial growth is possible under the conditions of $y = 0.2$ and $T = 740$ °C. That is, InN/$In_yGa_{1-y}N$ SLs ($y = 0.2$) can be fabricated under the relatively high growth temperatures, while it is difficult to grow lInN/nGaN SLs under the temperature range. These results imply that it is indispensable to consider the effect of lattice constraint to control the composition of coherently grown $In_xGa_{1-x}N$ on $In_yGa_{1-y}N$.

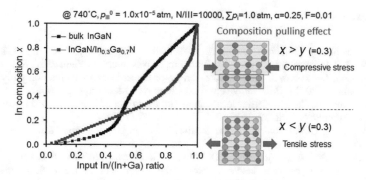

Fig. 6.13 In composition in InGaN as a function of input indium partial pressure ratio in p_{III}^0. Black and red squares indicate the calculation results of InGaN/In$_y$Ga$_{1-y}$N ($y = 0.3$) and bulk InGaN, respectively. The dotted line shows $x = 0.3$

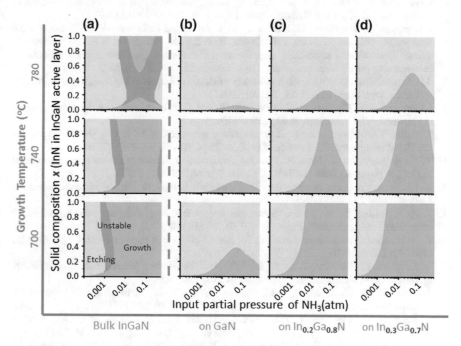

Fig. 6.14 Phase diagrams of InGaN MOVPE. **a** bulk InGaN, **b** InGaN on GaN, **c** InGaN on In$_y$Ga$_{1-y}$N ($y = 0.2$) and InGaN on In$_y$Ga$_{1-y}$N ($y = 0.3$). Blue, green and yellow regions indicate growth, unstable and etching conditions, respectively

References

1. S.F. Chichibu, A. Uedono, T. Onuma, B.A. Haskell, A. Chakraborty, T. Koyama, P.T. Fini, S. Keller, S.P. DenBaars, J.S. Speck, U.K. Mishra, S. Nakamura, S. Yamaguchi, S. Kamiyama, H. Amano, I. Akasaki, J. Han, T. Sota, Origin of defect-insensitive emission probability in In-containing (Al, In, Ga)N alloy semiconductors. Nat. Mater. **5**, 810 (2006)
2. P.R.C. Kent, A. Zunger, Carrier localization and the origin of luminescence in cubic InGaN alloys. Appl. Phys. Lett. **79**, 1977 (2001)
3. L.-W. Wang, Calculations of carrier localization in $In_xGa_{1-x}N$. Phys. Rev. B **63**, 245107 (2001)
4. T. Suski, T. Schulz, M. Albrecht, X.Q. Wang, I. Gorczyca, K. Skrobas, N.E. Christensen, A. Svane, The discrepancies between theory and experiment in the optical emission of monolayer In(Ga)N quantum wells revisited by transmission electron microscopy. Appl. Phys. Lett. **104**, 182103 (2014)
5. I. Gorczyca, T. Suski, N.E. Christensen, A. Svane, Band structure and quantum confined stark effect in InN/GaN superlattices. Cryst. Growth Des. **12**, 3521 (2012)
6. I. Gorczyca, T. Suski, N.E. Christensen, A. Svane, Hydrostatic pressure and strain effects in short period InN/GaN superlattices. Appl. Phys. Lett. **101**, 092104 (2012)
7. I. Gorczyca, A. Kamińska, G. Staszczak, R. Czernecki, S.P. Łepkowski, T. Suski, H.P.D. Schenk, M. Glauser, R. Butté, J.-F. Carlin, E. Feltin, N. Grandjean, N.E. Christensen, A. Svane, Anomalous composition dependence of the band gap pressure coefficients in In-containing nitride semiconductors. Phys. Rev. B **81**, 235206 (2010)
8. S.P. Łepkowski, I. Gorczyca, Ab initio study of elastic constants in $In_xGa_{1-x}N$ and $In_xAl_{1-x}N$ wurtzite alloys. Phys. Rev. B **83**, 203201 (2011)
9. I. Gorczyca, N.E. Christensen, E.L. Peltzery Blanca, C.O. Rodriguez, Optical phonon mades in GaN and AlN. Phys. Rev. B **51**, 11936 (1995)
10. T. Saito, Y. Arakawa, Atomic structure and phase stability of $In_xGa_{1-x}N$ random alloys calculated using a valence-force-field method. Phys. Rev. B **60**, 1701 (1999)
11. Y. Kangawa, T. Ito, A. Mori, A. Koukitu, Anomalous behavior of excess energy curves of $In_xGa_{1-x}N$ grown on GaN and InN. J. Cryst. Growth **220**, 401 (2000)
12. T. Ito, Recent progress in computer-aided materials design for compound semiconductors. J. Appl. Phys. **77**, 4845 (1995)
13. T. Ito, K.E. Khor, S. Das, Sarma, "Systematic approach to developing empirical potentials for compound semiconductors". Phys. Rev. B **41**, 3893 (1990)
14. T. Ito, Empirical interatomic potentials for nitride compound semiconductors. Jpn. J. Appl. Phys. **37**, L574 (1998)
15. Y. Kangawa, K. Kakimoto, T. Ito, A. Koukitu, Analysis of compositional instability of InGaN by Monte Carlo simulation. J. Cryst. Growth **298**, 190 (2007)
16. A. Koukitu, N. Takahashi, H. Seki, Thermodynamic study on metalorganic vapor-phase epitaxial growth of group III nitrides. Jpn. J. Appl. Phys. **36**, L1136 (1997)
17. A. Koukitu, N. Takahashi, T. Taki, H. Seki, Thermodynamic analysis of $In_xGa_{1-x}N$ alloy composition grown by metalorganic vapor phase epitaxy. Jpn. J. Appl. Phys. **35**, L673 (1996)
18. Y. Kawaguchi, M. Shimizu, M. Yamaguchi, K. Hiramatsu, N. Sawaki, W. Taki, H. Tsuda, N. Kuwano, K. Oki, T. Zheleva, R.F. Davis, The formation of crystalline defects and crystal growth mechanism in $In_xGa_{1-x}N$/GaN heterostructure grown by metalorganic vapor phase epitaxy. J. Cryst. Growth **189/190**, 24 (1998)

Chapter 7
Initial Epitaxial Growth Processes of III-Nitride Compounds

Toru Akiyama

The epitaxial growth of thin films is controlled by various growth processes that involve adsorption of atoms and molecules onto a reconstructed surface, their subsequent diffusion across the surface, dissociation of molecules, and desorption from the surface. In Chap. 4, we have studied that an ab initio-based approach incorporating the free energy of gas phase is useful to evaluate the influence of temperatures and gas-phase pressures on the stability of reconstructed surfaces of III-nitride compounds. This method is also applicable to investigate the growth kinetics and processes of epitaxial growth. Indeed, several ab initio calculations have revealed microscopic behaviors of adatoms such as adsorption, migration, and desorption on III-nitride surfaces [1–3]. Although these ab initio studies successfully elucidate some aspects in the growth processes of group III nitrides, their results are limited at 0 K without incorporating the growth parameters such as temperature and pressures. In this chapter, the feasibility of the chemical potential approach to the epitaxial growth processes, such as molecular beam epitaxy (MBE) growth and metal-organic vapor-phase epitaxy (MOVPE) growth of III-nitride compounds, are described. Furthermore, kinetic growth processes on these surfaces are also investigated using kinetic Monte Carlo (kMC) (See Sect. 2.3) and electron-counting Monte Carlo (ECMC) simulations [4, 5] on the basis of the results of adsorption, desorption, and migration energies obtained by ab initio calculations. The versatility of chemical potential approach is exemplified by the epitaxial growth of various III-nitride compounds including polar, nonpolar, and semipolar orientations. Several perspectives for more realistic simulations of crystal

T. Akiyama (✉)
Department of Physics Engineering, Mie University, Tsu, Japan
e-mail: akiyama@phen.mie-u.ac.jp

© Springer International Publishing AG, part of Springer Nature 2018
T. Matsuoka and Y. Kangawa (eds.), *Epitaxial Growth of III-Nitride Compounds*,
Springer Series in Materials Science 269,
https://doi.org/10.1007/978-3-319-76641-6_7

growth for various materials as functions of growth conditions, such as temperature and gas-phase pressure, are discussed.

7.1 Adatom Kinetics on AlN Polar Surfaces During MOVPE

The optimization of AlN growth condition could be achieved with an understanding of the growth mechanisms. However, the atomic-scale growth processes on AlN surfaces, such as adatom kinetics during epitaxial growth, still remain unclear. Theoretical studies on the properties and growth of AlN are sparse and have focused on the atomic and electronic structure [6–10]. In situ reflectance data of AlN films grown with MOVPE have reported that the growth rate on AlN(0001) surfaces significantly depends on the carrier gas species [11]. In addition, it has also been reported that the growth under N-rich conditions is much faster than that under H-rich conditions. These experimental finding can be interpreted using the super-saturation of Al, depending on the partial pressure of H_2, [12] but the effects of the growth processes on AlN(0001) surface during MOVPE growth have not been examined. Detailed studies for adatom kinetics on AlN surfaces would provide a deeper understanding of the growth processes. Indeed, calculations for the adsorption and diffusion behaviors of Al and N atoms on technologically relevant AlN(0001) surfaces during MOVPE growth have revealed that the surface reconstruction crucially affects the adatom kinetics [13].

On the basis of the calculated surface structures shown in Figs. 4.23a and 4.24a, the kinetics of Al and N adatoms on the AlN(0001) surface can be clarified. The potential energy surface (PES) calculations for an N adatom on the surface with another N adatom and with H atoms show that the most stable adsorption site for the additional N atom is located near the pre-adsorbed N atom. However, the adsorption energies (−2.29 eV) indicate that the N atom desorption occurs even at 0 K. This finding thus suggests that the adsorption of Al adatoms that attach to the outermost N atom is necessary to form AlN layers on AlN(0001) surfaces. These results are contradictory to the adatom kinetics on a GaN(0001) surface, in which N-rich surface morphology can be kinetically stabilized [1]. The PES calculations for an Al adatom on the surface suggest that the Al atom adsorption behavior significantly depends on the reconstruction [13]. Figure 7.1 shows the PES of an additional Al atom on surfaces with an N adatom and H atom, respectively. As shown in Fig. 7.1a, the most stable adsorption site on the surface under low H_2 pressure conditions is located above the N adatom, and a strong Al–N bond with a bond length of 1.81 Å is formed, which is similar to the bond length in bulk AlN (1.91 Å). The Al–N bond formation results in an adsorption energy of $E_{ad} = -3.25$ eV, which is much lower than the adsorption energy at high H_2 pressure conditions. To move adjacent stable adsorption sites, there is a transition state for diffusion. The Al adatom positions in the transition state are located close to the topmost Al atom, which does not have an Al–N bond with the N adatom.

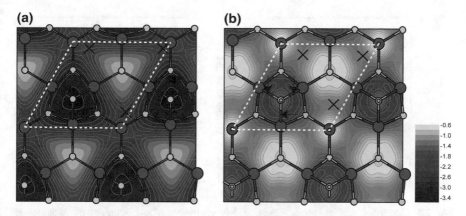

Fig. 7.1 Contour plots of the PES for an Al adatom on the reconstructed AlN(0001) surfaces under **a** H-poor (N_{ad} in Fig. 4.23a) and **b** H-rich ($N_{ad} - H + Al-H$ in Fig. 4.23a) conditions. Large, small, and tiny circles represent Al, N, and H atoms, respectively. Each contour line in **a** and **b** represents an energy step of 0.1 and 0.2 eV, respectively. Dashed rectangles denote the surface unit cells. Arrows and crosses in the unit cell represent minima and saddle points of the PES, respectively. Reproduced with permission from Akiyama et al. [13]. Copyright (2012) by American Institute Physics

The energy barrier for migration is 0.81 eV. This transition state corresponds to the dissociation of an Al–N bond between the N adatom and the topmost Al atom.

In contrast, the most stable surface adsorption sites under high H_2 pressure conditions are located between topmost Al–N bonds, as shown in Fig. 7.1b. An Al–Al bond (bond length 2.66 Å) is formed between the Al adatom and topmost surface Al atom. The adsorption energy is $E_{ad} = -1.89$ eV, corresponding to the energy required to form an Al–Al bond. The energetically lowest transition sites for diffusion are located close to the hexagonal sites without an N adatom, and the corresponding energy barrier E_{diff} is 0.75 eV. The physical origin of the energy barrier is attributed to the formation of a weak Al–Al bond. This bond is stretched by 24% from the original Al–Al bond, indicating a significantly reduced bond strength. Despite the small energy barrier difference, which depends on the growth conditions, the physical origin of the energy barrier is quite different.

The adsorption behavior of Al adatoms on the reconstructed AlN(0001) surface under realistic condition can be evaluated by the surface phase diagrams as functions of temperature and pressure. Figure 7.2 shows the calculated surface phase diagrams of AlN(0001) as functions of temperature and Al pressure. The surface phase diagrams demonstrate that the surface structure depending on H_2 pressure crucially affects the adsorption behavior of Al adatoms. Owing to the low adsorption energy ($E_{ad} = -3.25$ eV), the temperature for the adsorption of Al adatoms under H-poor condition ranges from 860 to 1220 K depending on Al pressure, as shown in Fig. 7.2a. In contrast, the temperature for the adsorption under H-rich condition shown in Fig. 7.2b is ranging from 520 to 740 K. This is because the adsorption energy $E_{ad} = -1.89$ eV under H-rich condition is much

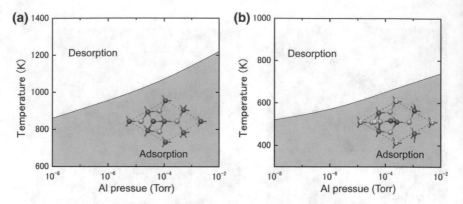

Fig. 7.2 Calculated surface phase diagrams for adsorption of Al adatoms as functions of temperature and Al pressure on the reconstructed AlN(0001) surface under **a** H-poor (N_{ad} in Fig. 4.23a) and **b** H-rich (N_{ad}–H + Al–H in Fig. 4.23a) conditions. In the equilibrium state, the adsorption of Al adatoms occurs in the regions emphasized by the shaded area. Top views of surface structures are also shown

higher than the gas-phase chemical potential of Al atoms in (2.63) around 1370 K. The surface phase diagram shown in Fig. 7.2b thus suggests that the additional Al atoms easily desorb from the surface under H-rich condition.

Furthermore, we can deduce the adatom kinetics more quantitatively using the calculated adsorption energies and diffusion barriers. This can be accomplished by estimating the diffusion length, L_{diff}, expressed as

$$L_{diff} = \sqrt{2}D\tau, \tag{7.1}$$

where D is the diffusion coefficient and τ is the lifetime of the Al adatom between the adsorption and desorption events. In the cases of AlN(0001) surfaces, the difference in E_{diff}, which is within 0.06 eV depending on the reconstruction, does not contribute to the difference in diffusion length because D is proportional to $\exp(-E_{diff}/k_BT)$. On the contrary, the desorption probability defined by $1/\tau$ is proportional to $\exp(-E_{de}/k_BT)$, where E_{de} is the desorption energy. The desorption energy of an Al adatom, depending on the growth conditions, thus affects the diffusion length. Figure 7.3 shows τ and L_{diff} of an Al adatom under low and high H_2 pressure conditions (at an Al pressure of 1×10^{-3} Torr) using the kMC simulations described in Sect. 2.3. The estimated τ and L_{diff} for low H_2 pressure conditions shown in Fig. 7.3a, b are four and two orders of magnitude larger than for high H_2 pressure conditions, respectively. This indicates that the growth under N-rich conditions is much faster than that under H-rich conditions. Although the adsorption processes for a monolayer AlN film should be verified to obtain the growth rate more quantitatively, this conclusion is qualitatively consistent with recent in situ reflectance data of AlN films grown with MOPVE [11].

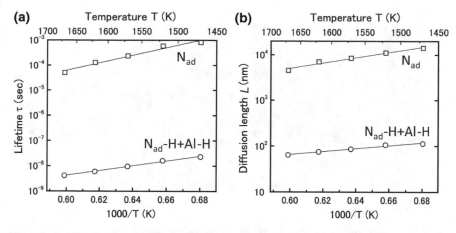

Fig. 7.3 Calculated **a** lifetime τ and **b** diffusion length L_{diff} of Al adatoms at Al pressure of 1.0×10^{-3} Torr on the reconstructed AlN(0001) surfaces under H-rich (N_{ad}–H + Al–H in Fig. 4.23a) and H-poor (N_{ad} in Fig. 4.23a) conditions as a function of reciprocal temperature obtained by kMC simulations. The growth temperature in 11 (1370 K) is shown by green lines

7.2   Adatom Kinetics on Semipolar AlN($11\bar{2}2$) Surface During MOVPE

It has been recently reported that the epitaxial growth of AlN strongly depends on hydrogen ambient in the MOVPE [11, 14]. More importantly, the growth of AlN on a semipolar AlN($11\bar{2}2$) surface is sensitive to H_2 pressure [14]. The growth rate decreases with increasing H_2 pressure, similarly to the case of AlN(0001) surface [11]. Despite these experimental studies, there have been few studies on the growth of AlN on semipolar orientations. Although the optimization of growth conditions could be achieved by understanding the growth mechanisms, the atom-scale growth mechanisms on semipolar AlN surfaces remain unclear. Detailed theoretical studies on the adatom kinetics on AlN($11\bar{2}2$) surface, which would provide deeper understanding of the growth processes, are still lacking.

On the basis of the calculated surface structures in Sect. 4.3, the adsorption behavior of Al atoms on the AlN($11\bar{2}2$) surface can be clarified. Indeed, the calculations of the PES for Al atoms on the c(2×2) surfaces indicate that the adsorption behavior of the Al atoms significantly depends on the type of reconstruction [15]. Figure 7.4 illustrates the PES of the additional Al atoms on the c (2×2) surfaces. The most stable adsorption site on the surface under H-rich conditions (Al$_{ad}$–H + N$_{ad}$–H + AlH$_2$ shown in Figs. 4.23f and 4.24 f) is located above the topmost N atom (arrows in Fig. 7.4a). This results in the formation of an Al–N bond (bond length is 1.86 Å) between the Al atom and the topmost N atom. The adsorption energy is −2.15 eV, originating from the energy gain to form an Al–N bond. This value corresponds to the desorption temperatures ranging from

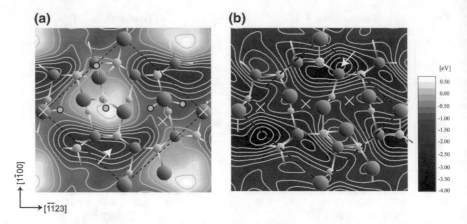

Fig. 7.4 Contour plots of the PES for an Al adatom on the reconstructed AlN($11\bar{2}2$) surfaces under **a** H-rich (Al$_{ad}$–H + N$_{ad}$–H + AlH$_2$ shown in Fig. 4.23f) and **b** H-poor (Al$_{ad}$ shown in Fig. 4.23f) conditions. Large, small, and tiny circles represent Al, N, and H atoms, respectively. Each contour line represents an energy step of 0.25 eV. Dashed rectangles denote the surface unit cells. Arrows and crosses in the unit cell represent minima and saddle points of the PES, respectively. Reproduced with permission from Akiyama et al. [15]. Copyright (2015) by the Japan Society of Applied Physics

700 and 900 K, which are estimated by comparing the calculated adsorption energy of the Al atom with the gas-phase chemical potential of the Al atom obtained using (2.63), indicating that most of the additional Al atoms easily desorb from the surface under H-rich conditions. The energetically lowest transition sites for the migration of Al atoms are located above the topmost Al atoms (crosses in Fig. 7.4a) and the energy barrier for migration E_{diff} is 1.5 eV. The physical origin of the energy barrier is attributed to the dissociation of the Al–N bond between the Al atom and the topmost N atoms.

In contrast to the case of Al$_{ad}$–H + N$_{ad}$–H + AlH$_2$, the most stable adsorption site on the surface under H-poor conditions (Al$_{ad}$ shown in Figs. 4.23f and 4.24f) is located close to the Al-lattice site above the topmost N atom (arrow in Fig. 7.4b) and strong Al–N bonds whose lengths (1.97–2.02 Å) are similar to the bond length in bulk AlN (1.91 Å) are formed. The formation of two Al–N bonds produces the adsorption energy of −3.91 eV, which is much lower than that under H-rich conditions. This low adsorption energy corresponds to the desorption temperatures ranging from 1200 and 1500 K depending on Al pressure. It is thus likely that around 1400 K, most of the additional Al atoms stably adsorb on the surface under H-poor conditions. The transition states for the migration are located between the topmost Al atoms (crosses in Fig. 7.4b) and the energy barrier E_{diff} is 1.6 eV. These transition states correspond to the dissociation of Al–N bonds and the formation of weak Al–Al bonds between the Al atom and the topmost Al atoms (bond length is ~ 2.8 Å) during the migration. In spite of the small difference between the energy barrier under H-poor conditions and that under H-rich conditions, the

Table 7.1 Calculated adsorption energy E_{ad} at the most stable site and energy barrier for migration E_{diff} of Al atoms on AlN($11\bar{2}2$) surface along the $[1\bar{1}00]$ direction under H-rich and H-poor (Al$_{ad}$–H + N$_{ad}$–H + AlH$_2$ and Al$_{ad}$ shown in Fig. 4.23f, respectively) conditions. Values in parentheses for the AlN($11\bar{2}2$) surface are E_{diff} values obtained along the $[\bar{1}\bar{1}23]$ direction. The calculated values on the AlN(0001) surface shown in Fig. 7.1 are also shown

Orientation	Reconstruction	E_{ad} (eV)	E_{diff} (eV)
AlN($11\bar{2}2$)	Al$_{ad}$–H + N$_{ad}$–H + AlH$_2$	−2.15	1.5 (1.6)
	Al$_{ad}$	−3.91	1.6 (1.9)
AlN(0001)	N$_{ad}$–H + Al–H	−1.89	0.75
	N$_{ad}$	−3.25	0.81

physical origins of the energy barriers are different from each other. It should be noted that the adsorption behavior of the Al atoms on the AlN($11\bar{2}2$) surface under H-poor conditions is similar to that on the GaN($11\bar{2}2$) surface [16].

Table 7.1 shows the calculated adsorption energy (E_{ad}) and energy barrier for the migration (E_{diff}) of Al atoms on AlN($11\bar{2}2$) surface, along with those on AlN (0001) surface discussed in Sect. 7.1. Trends in E_{ad} and E_{diff} depending on the type of reconstruction on the AlN($11\bar{2}2$) surface are similar to those on the AlN(0001) surface: The adsorption energy under H-poor conditions is lower than that under H-rich conditions, and there is a small energy difference in E_{diff} depending on the type of surface reconstruction. Owing to the difference in atomic arrangement depending on the surface orientation, the energy barrier for migration on AlN($11\bar{2}2$) surface (E_{diff} = 1.5–1.6 eV) is higher than that on the AlN(0001) surface (E_{diff} = 0.75–0.81 eV). It should also be noted that there is a difference in E_{diff} depending on the direction of migration. This implies an anisotropy in the migration behavior on the AlN($11\bar{2}2$) surface, as seen on other nonpolar and semiopolar surfaces.

Furthermore, the adsorption behavior of Al adatoms on the reconstructed AlN($11\bar{2}2$) surface during the MOVPE can be evaluated by the surface phase diagrams as functions of temperature and pressure. Figure 7.5 shows the calculated surface phase diagrams of AlN($11\bar{2}2$) as functions of temperature and Al pressure. The surface phase diagrams demonstrate that the surface structure depending on H$_2$ pressure crucially affects the adsorption behavior of Al adatoms. The temperature for the adsorption under H-rich condition shown in Fig. 7.5a is 600–800 K. This is because the adsorption energy E_{ad} = −2.15 eV under H-rich condition is much higher than the gas-phase chemical potential of Al atoms in (2.63) around 1370 K. The surface phase diagram shown in Fig. 7.5a thus suggests that the additional Al atoms easily desorb from the surface under H-rich condition. In contrast, owing to the low adsorption energy (E_{ad} = −3.91 eV), the temperature for the adsorption of Al adatoms under H-poor condition ranges from 1030 to 1450 K depending on Al pressure, as shown in Fig. 7.5b. It is thus likely that around 1370 K, most of the additional Al atoms stably adsorb on the surface under H-poor condition.

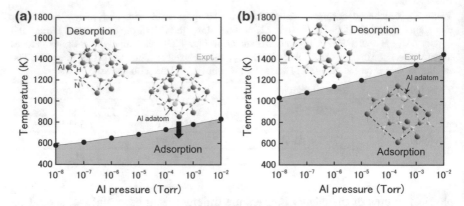

Fig. 7.5 Calculated surface phase diagrams for adsorption of Al adatoms as functions of temperature and Al pressure on the reconstructed AlN($11\bar{2}2$) surface under **a** H-rich (Al_{ad}–H + N_{ad}–H + AlH$_2$ shown in 4.23f) and **b** H-poor (Al_{ad} shown in Fig. 4.23f) conditions. In the equilibrium state, the adsorption of Al adatoms occurs in the regions emphasized by the shaded area. Top views of surface structures are also shown. The growth temperature in 15 (1370 K) is shown by green lines. Reproduced with permission from Akiyama et al. [17]. Copyright (2016) by the Japan Society of Applied Physics

By using the calculated PES, numerical analysis of adatom kinetics on the surface using the kMC simulations has been performed [17]. Figure 7.6 shows the calculated surface lifetime τ and the diffusion length L_{diff} of Al adatoms on AlN($11\bar{2}2$) surfaces as a function of reciprocal temperature. Owing to the adsorption energy difference between Al_{ad} and Al_{ad}–H + N_{ad}–H + AlH$_2$, the calculated surface lifetime under H-rich condition shown in Fig. 7.6a is four orders of magnitude smaller than that under H-poor condition. The calculated diffusion length under H-poor condition Fig. 7.6b is found to be four orders of magnitude larger than that under H-rich condition. This is because the adsorption energy under H-poor condition (−3.91 eV) is much lower than that under H-rich condition (−2.15 eV). Since the migration probability is proportional to $\exp(-E_{diff}/k_BT)$, as shown in (2.64), the difference in E_{diff} within 0.1 eV between Al_{ad} and Al_{ad}–H + N_{ad}–H + AlH$_2$ hardly contributes to the difference in the migration behavior of Al adatoms. It is thus implied that the AlN($11\bar{2}2$) surface under H-poor condition grows more rapidly than that under H-rich condition because of the higher sticking probability for Al adsorption under H-poor condition. The insights derived from our calculations are qualitatively consistent with experimental observations [14].

It should be noted that the hydrogen pressure dependence during initial growth processes on the AlN($11\bar{2}2$) surface is similar to that on the AlN(0001) surface: The growth on the AlN(0001) surface under H-poor condition is much faster than that under H-rich condition [11]. Since the adsorption energy on the AlN($11\bar{2}2$) surface under H-poor (H-rich) condition is 0.66 (0.26) eV lower than that on the AlN(0001) surface, the calculated surface lifetime on the AlN($11\bar{2}2$) surface becomes large compared with that on the AlN(0001) surface. On the other hand, the diffusion

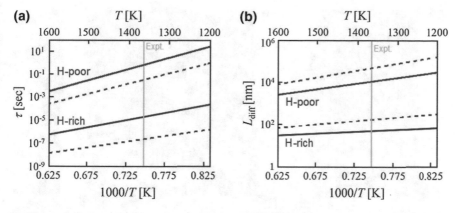

Fig. 7.6 Calculated **a** lifetime τ and **b** diffusion length L_{diff} of Al adatoms at Al pressure of 1.0×10^{-3} Torr on the reconstructed AlN($11\bar{2}2$) surfaces (solid lines) **a** H-poor (Al$_{ad}$ shown in 4.23f) and **b** H-rich (Al$_{ad}$–H + N$_{ad}$–H + AlH$_2$ in Fig. 4.23f) conditions as a function of reciprocal temperature obtained by kMC simulations. The growth temperature in 15 (1370 K) is shown by green lines. The calculated results on the reconstructed AlN(0001) surfaces are also shown by dashed lines. Reproduced with permission from Akiyama et al. [17]. Copyright (2016) by the Japan Society of Applied Physics

length on the AlN($11\bar{2}2$) surface is smaller than that on the AlN(0001) surface. The difference in diffusion length between the AlN(0001) and ($11\bar{2}2$) surfaces originates from the difference in the energy barrier for migration. The energy barrier on the AlN($11\bar{2}2$) surface is ~ 1.6 eV, while that on the AlN(0001) surface is less than 0.81 eV.

7.3 Adatom Kinetics on Semipolar AlN($1\bar{1}01$) and ($1\bar{1}02$) Surfaces During MOVPE

It has been recently reported that the growth of AlN on semipolar ($1\bar{1}02$) orientations is sensitive to H$_2$ pressure: High-density pits consisting of AlN($1\bar{1}01$) surface are formed under relatively low H$_2$ pressure conditions while smooth AlN($1\bar{1}02$) surface is obtained under high H$_2$ pressure conditions at initial growth processes. Furthermore, high-density pits perfectly disappear if high H$_2$ pressure is changed to low H$_2$ pressure after the growth of AlN($1\bar{1}02$) epi-layer whose thickness is 100 nm [18]. Indeed, the calculations for the growth processes on AlN($1\bar{1}02$) and ($1\bar{1}01$) surfaces during MOVPE growth have suggested that the surface orientation crucially affects the adatom kinetics.

In order to analyze the growth processes on semipolar orientations, the adsorption behavior of Al adatom on AlN($1\bar{1}02$) and AlN($1\bar{1}01$) surfaces has been investigated [19]. Figure 7.7 shows the PES of Al adatom on AlN($1\bar{1}02$) and

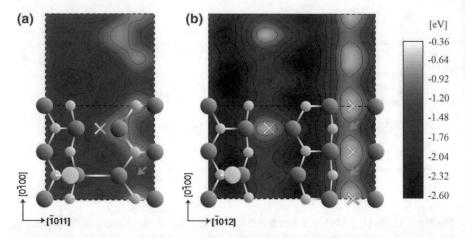

Fig. 7.7 Contour plots of the PES for an Al adatom for **a** on AlN($1\bar{1}02$) and **b** AlN($1\bar{1}01$) surfaces under MOVPE conditions. Top views of the surface unit cell with Al adatom located at the most stable site are also shown by dashed rectangles. Large and small circles represent Al and N atoms, respectively. Crosses denote H atoms attaching on the surface. Arrows denote the saddle points of PES

($1\bar{1}01$) surfaces. The calculated PES demonstrate that adsorption behavior of Al adatom strongly depends on the surface reconstruction. As shown in Fig. 7.7a, the most stable adsorption site on AlN($1\bar{1}02$) surface is located above topmost N atom. The adsorption energy (-1.89 eV) originates from large number of excess electrons caused by Al adatom. Similar to AlN($1\bar{1}02$) surface, the most stable adsorption site on AlN($1\bar{1}01$) surface is positioned above topmost N atom and single Al–N bond is formed. The number of excess electron on AlN($1\bar{1}01$) surface is smaller than that on AlN($1\bar{1}02$) surface. As a result, the adsorption energy (-2.49 eV) on AlN($1\bar{1}01$) surface is slightly lower than that on AlN($1\bar{1}02$) surface.

Another important feature of the PES is the kinetic behaviors of Al adatom on the surface that are also crucial for the epitaxial growth. The PES demonstrate that the energy barriers for diffusion of Al adatom strongly depend on surface orientations. The energies of adsorption sites on AlN($1\bar{1}02$) surface are relatively higher than those on AlN($1\bar{1}01$) surface. Furthermore, there are high potential areas along the [0$\bar{1}$00] direction on both surface orientations. In particular, the energy barrier for AlN($1\bar{1}02$) surface along the [$\bar{1}011$] direction (0.8 eV) is lower than that for AlN($1\bar{1}01$) surface along the [$\bar{1}012$] direction (1.4 eV). The difference in energy barriers can be interpreted in terms of the bond formation at the transition state structures. The lateral position of Al adatom in the transition state on AlN is located between topmost Al and N atoms (arrow in Fig. 7.7a). This transition state corresponds to the formation of stretched Al–N bond (2.24 Å), as shown in Fig. 7.7a. On the other hand, the Al adatom in the transition state on AlN($1\bar{1}01$) surface is located near topmost H atoms (arrows in Fig. 7.7b). Two weak Al–H bonds are formed in the transition state structures on AlN as shown in Fig. 7.7b. The difference of energy barrier between AlN($1\bar{1}02$) and ($1\bar{1}01$) surfaces is reflected by the difference of Al adsorption energy at the local energy minima between AlN($1\bar{1}02$) and ($1\bar{1}01$) surfaces. It should be noted that there are high adsorption energy areas along the [$0\bar{1}00$] direction for AlN($1\bar{1}01$) surface due to the presence of H atoms. This high-energy region leads to anisotropy in diffusion behavior of Al adatom on AlN($1\bar{1}01$) surface.

On the basis of the calculated PES, numerical analysis of adatom kinetics on the surface has been performed using the kMC simulations described in Sect. 2.3 [19]. Figure 7.8 shows the calculated diffusion length L_{diff} and surface lifetime τ of Al adatom on AlN($1\bar{1}02$) and AlN($1\bar{1}01$) surfaces in typical growth temperature range. The calculated diffusion length L_{diff} on AlN($1\bar{1}02$) surface shown in Fig. 7.8a is an order of magnitude larger than that on AlN($1\bar{1}01$) surface. This is because the energy barrier for diffusion on AlN($1\bar{1}02$) surface is lower than that on AlN($1\bar{1}01$) surface. It is suggested that Al adatoms easily migrate on AlN($1\bar{1}02$) surface compared with that on AlN($1\bar{1}01$) surface. On the other hand, due to the adsorption energy difference between AlN($1\bar{1}02$) and ($1\bar{1}01$) surfaces, the calculated surface lifetime τ on AlN($1\bar{1}02$) surface shown in Fig. 7.8b is two orders of magnitude smaller than that on AlN($1\bar{1}01$) surface. This implies that AlN($1\bar{1}01$) surface grows more rapidly than AlN($1\bar{1}02$) surface because of the higher sticking probability for Al adsorption on the AlN($1\bar{1}01$) surface. Furthermore, the longer Al diffusion length on AlN($1\bar{1}02$) surface than that on AlN($1\bar{1}01$) surface indicates that AlN($1\bar{1}02$) surface is more smooth than AlN($1\bar{1}01$) surface. The insights derived from these calculations are qualitatively consistent with experimental observations [18].

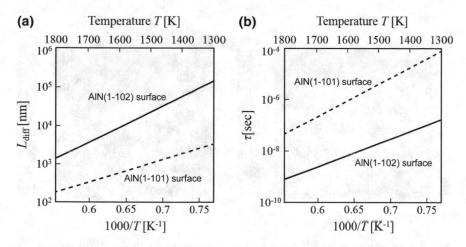

Fig. 7.8 Calculated **a** diffusion length L_{diff} and **b** lifetime τ of Al adatoms at Al pressure of 1.0×10^{-3} Torr on the reconstructed AlN($1\bar{1}02$) and **(b)** AlN($1\bar{1}01$) surfaces under MOVPE conditions as a function of reciprocal temperature obtained by kMC simulations. Solid and dashes lines represent the results on AlN($1\bar{1}02$) and ($1\bar{1}01$) surfaces, respectively

7.4   Adsorption Behavior of Al and N Atoms on Nonpolar 4H–SiC(11$\bar{2}$2) Surface

For III-nitride compounds, heterojunctions between nitrides and SiC substrate have attracted considerable interest, and abrupt AlN/SiC heterojunctions with a low density of dislocations have been successfully fabricated owing to the negligible lattice mismatch between SiC and AlN (less than 1%). Recently, the growth of 4H–AlN on 4H–SiC substrate has been performed [20] and very high-quality 4H–AlN can be realized by precise tuning of V/III ratio and SiC surface treatment using plasma assisted MBE [21, 22]. This implies that the influence of the V/III ratio on polytype of AlN(11$\bar{2}$0) grown on 4H–SiC(11$\bar{2}$0) substrate. For N-rich condition 2H–AlN films are grown on the substrate while 4H–AlN films are formed under Al-rich condition. These experimental findings indicate that the growth mechanisms of AlN films on the 4H–SiC substrate are influenced on the V/III ratio. However, the growth processes such as adsorption-desorption behavior of Al and N atoms and their surface migrations on 4H–SiC(11$\bar{2}$0) surface have been rarely investigated.

Figure 7.9 shows the calculated contour plots of the PES for Al and N adatoms on the ideal 4H–SiC(11$\bar{2}$0) surface [23]. Since the ideal 4H–SiC(11$\bar{2}$0) surface satisfies the electron counting (EC) rule [24] by considering electron transfer from Si dangling bonds to C dangling bonds, the ideal surface can be considered as a stable surface structure on nonpolar orientation. Indeed, the electronic states caused by C dangling bond are fully occupied and those by Si dangling bond are empty. As shown in Fig. 7.9a, the energetically lowest position for the Al atom is located close to the lattice site of C atom. On the other hand, the N atom has the lowest energy

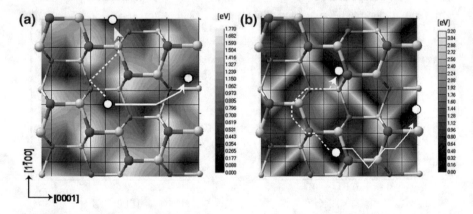

Fig. 7.9 Contour plots of the PES for an **a** Al and **b** N adatoms on 4H–SiC(11$\bar{2}$0) surfaces. The origin of energy is set to the total energy of the most stable site. Small grey and while circles denote Si and C atoms, respectively. Al/N adatom is represented by large circles. Reproduced with permission from Ito et al. [23]. Copyright (2009) by Elsevier

when it is located at the interstitial site above the top C atom (Fig. 7.9b). This implies that stable adsorption sites for Al and N atoms are different. These stable sites can be explained in terms of the EC rule [24]. For 4H–SiC($11\bar{2}0$) surface without adsorbents, all dangling bonds of Si atoms are empty and those of C atoms are fully occupied by electrons. Owing to the transfer of electrons, the surface becomes semiconducting and stabilized. When the Al atom adsorbs at the stable site, it forms two Al–C bonds and an Al–Si bond with its nearest neighbor C and Si atoms, respectively. Thus, the Al atom forms in total three covalent bonds whereas only less than two bonds are formed on the other adsorption site. Due to the formation of three covalent bonds, the number of excess electrons becomes only one in the unit cell. However, this adsorption site (where two C–N bonds and a Si–N bond are formed) is not the stable site for the N atom because the N atom can form two Si–N bonds and a C–N bond at the interstitial site. Since the bond energy of Si–N is small compared to that of C–N, [25] larger number of Si–N bonds causes the stabilization of the N atom. Table 7.2 shows the calculated adsorption energy for the Al and N atoms. The adsorption energy of the Al atoms (-6.71 eV) is found to be lower than that of the N atoms (-4.06 eV). This implies that the adsorption of Al atoms is more favorable than that of N atoms.

From Fig. 7.9, we can also deduce the migration behavior of adsorbed Al and N atoms. For the Al atom, symmetrically equivalent stable sites are placed with 5.3 Å interval. If we assume that the migration occurs by taking the migration pathways between nearest-neighbor stable sites, the migration energy barrier corresponds to the highest total energy position along the ridge between nearest neighbor-stable sites. Table 7.2 also shows the energy barriers based on this assumption. We obtain the values of 1.11 and 0.92 eV along the $[1\bar{1}00]$ and $[0001]$ directions, respectively. It seems that the migration along the $[0001]$ direction (arrow in Fig. 7.9a) is slightly easier than that along the $[1\bar{1}00]$ direction (dashed arrow in Fig. 7.9a). The low migration energy barrier along the $[0001]$ direction for the Al atom can be understood by the shorter distance between stable or metastable positions along the migration pathways. For instance, the distances along the $[0001]$ direction for the Al (N) atom are 1.88 (2.63) Å whereas those along the $[1\bar{1}00]$ direction are 5.26 (3.01) Å. For the N atom, the migration barriers are higher than those for Al atoms. Again, the migration along the $[0001]$ direction of N atoms (arrow in Fig. 7.9b) is easier than that along the $[1\bar{1}00]$ direction (dashed arrow in Fig. 7.9b). Since the C–N and Si–N bonds are short (1.45 and 1.83 Å, respectively) compared to Al–C

Table 7.2 Calculated adsorption energy E_{ad} at the most stable site and energy barrier for migration E_{diff} along the $[0001]$ and $[1\bar{1}00]$ directions for Al and N atoms on 4H–SiC($11\bar{2}0$) surface. The unit of energy is eV

	Al	N
E_{ad}	−6.71	−4.06
E_{diff} along $[0001]$	0.92	1.59
E_{diff} along $[1\bar{1}00]$	1.11	1.95

and Al–C bonds (2.07 and 2.45 Å), breaking of the C–N bond occurs at the transition state for the migration along the [1$\bar{1}$00] direction. This migration corresponds to a jumping process through the trench located between top surfaces. In contrast, the migrations along the [0001] direction coincide with the atomic arrangement of the top surface. Along this direction, breaking and formation of Si–N and C–N bonds occur simultaneously, resulting in lower migration barrier.

The adsorption-desorption behavior under the MBE conditions is also clarified using the surface phase diagrams of Al and N atoms on the 4H–SiC(11$\bar{2}$0) surface. Figures 7.10a, b shows the calculated phase diagram for the adsorption of Al and N atoms as functions of temperature and Al- and N-beam equivalent pressure (BEP), respectively. As shown in Fig. 7.10a, the desorption temperature of the Al atom changes from 1800 to 2370 K depending on Al–BEP. These values are much higher than the experimental conditions (1223 K). Furthermore, even in the metastable sites, the desorption temperature for surface phase transition (1750–2300 K) is much higher than the experimentally reported temperature. From these findings, we can deduce that the Al atom adsorbs at every adsorption site and migrate without desorption. On the other hand, the desorption temperature of the N atom which is located at the most stable site changes from 1100 to 1500 K. By comparing the calculated temperature range for adsorption with the experiment, the N atom adsorbs on the stable site when N–BEP is larger than 1×10^{-6} Torr. However, the desorption temperature for the N atom located at metastable sites becomes lower than that at the most stable site. As a result, the N atom located at metastable sites desorbs over the wide range of N–BEP. Therefore, the N atom can adsorb only at the most stable site and is desorbed during its migration toward other adsorption sites. These results thus indicate that the adsorption-desorption behaviors of Al and N atoms are quite different with each other.

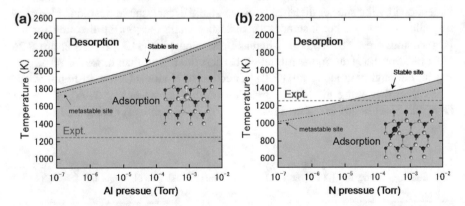

Fig. 7.10 Calculated surface phase diagrams for adsorption for **a** Al and **b** N adatoms as functions of temperature and pressure at the most stable site (solid line) and metastable sites (dotted line) on 4H–SiC(11$\bar{2}$0) surfaces along with atomic structures. Dashed line represents the growth temperature (1225 K) of AlN films on 4H–SiC(11$\bar{2}$0) surfaces in [21, 22]

Based on the calculated results, we can furthermore conjecture the growth processes of AlN films with different crystal structures depending on the V/III ratio [21, 22]. If we assume that the growth under Al-rich condition is dominated by Al atoms, this growth can be interpreted in terms of the adsorption behavior of Al atoms. Since the Al atom adsorbs close to the lattice site of 4H–SiC, it is likely that the growing AlN films inherit the atomic arrangement of the substrate, resulting in the formation of 4H–AlN. For N-rich condition, in contrast, we can consider that the growth is dominated by N atoms. Since the N atom cannot adsorb at the lattice site of the substrate but is incorporated at the interstitial site, the growing AlN films are likely not to inherit the atomic arrangement of the substrate and then the most stable crystal structure (2H–AlN) can be formed. Although details of the AlN growth processes on SiC($11\bar{2}0$) substrate should be verified more quantitatively, this is reasonably consistent with the experimental fact that 2H- and 4H–AlN films are grown under N- and Al-rich conditions, respectively [21, 22].

Figure 7.11 shows the calculated contour plots of the PES for Al and N on the 4H–SiC($11\bar{2}0$) surface after N and Al pre-depositions, respectively [26]. This figure shows the distinctive atomic arrangements of the Al and the N atoms simultaneously located on the surface. After the N deposition, the Al adatom independently occupies the lattice sites near the 4H structure (hereafter the 4H-sites) similar to that for the single Al adatom on the surface shown in Fig. 7.9a. The adsorption energy of Al adatom after N pre-deposition (-7.08 eV) is similar to that of single Al adatom. In contrast, the N adatom stably forms the dimer structure with the pre-deposited Al atom, resulting in lower adsorption energy of -5.94 eV. Figure 7.12 shows the calculated surface phase diagrams for the adsorption. This figure suggest that the N adatom stably forms the dimer structure even at $T > 1200$ K in contrast to the results for the single N adatom shown in Fig. 7.10b [23]. This is due to the fact that the Al pre-deposition lowers the adsorption energy for the N

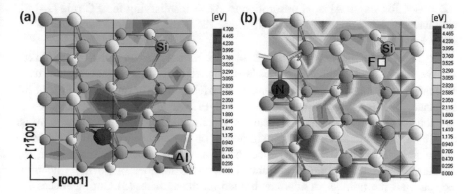

Fig. 7.11 Contour plots of the PES for an **a** Al and **b** N adatoms on 4H–SiC($11\bar{2}0$) surfaces after N and Al adsorptions, respectively. The origin of energy is set to the total energy of the most stable site. Small grey and while circles denote Si and C atoms, respectively. Al/N adatom is represented by large circles

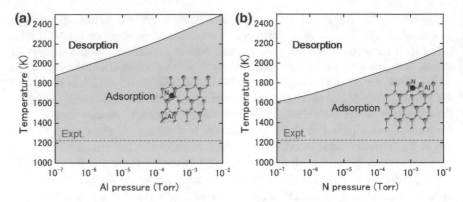

Fig. 7.12 Calculated surface phase diagrams for adsorption for **a** Al and **b** N adatoms as functions of temperature and pressure on 4H–SiC(11$\bar{2}$0) surfaces after N and Al adsorptions, respectively. The atomic structures are also shown. Dashed line represents the growth temperature (1225 K) of AlN films on 4H–SiC(11$\bar{2}$0) surfaces in [21, 22]

adatom (−5.94 eV). Furthermore, it is clear that the Al adatom can diffuse across the surface to favor the 4H-sites even after the N adsorption, since the migration barriers for the Al adatom are small such as 1.44 and 1.78 eV along the [0001] and the [1$\bar{1}$00], respectively. On the other hand, as shown in Fig. 7.11b, the N adatom stably form the dimer structure with the predeposited Al different from the 4H atomic arrangements on the surface. It should be noted that the metastable site for the N adatom is found at the F site corresponding to the 2H-site denoted by the open square in Fig. 7.11b. This suggests that Al deposition stabilizes the N adatom to form the Al–N dimer structure or to occupy the 2H site (F site) on the 4H–SiC(11$\bar{2}$0) surface. According to these results, the stable atomic arrangements appear to lower the number of electrons in the surface dangling bonds from $\Delta Z = +1$ for single Al or N adsorption to $\Delta Z = 0$ satisfying the EC rule [24] for simultaneous adsorption of Al and N, where ΔZ is the number of electrons in the surface dangling bonds.

To investigate the growth process of AlN monolayer on the 4H–SiC(11$\bar{2}$0) surface, the ECMC method [4, 5] has been applied [26]. This method enables us to evaluate the resultant atomic arrangements of AlN monolayer on the 4H–SiC(11$\bar{2}$0) substrate. The following assumptions in the ECMC simulations are considered. (1) The adsorption of the atoms occurs at specific surface sites. (2) Depending on the III/V ratio, Al and N adatoms are impinged on the surface to migrate their stable lattice sites. To keep the same number of Al and N atoms in the monolayer, the excess Al or N adsorption is prohibited when the number of Al or N adatoms exceeds half the number of adatoms limited on the surface. (3) During successive adsorption, an exchange process occurs between adatoms and vacant sites on the same plane. (4) Dimerization and nearest-neighbor exchange between adatoms are also included at each MC step in the equilibration procedure. (5) The system energy E (in eV/atom) is given by

$$E = E_{bond} + 0.2 \Delta Z, \tag{7.2}$$

where E_{bond} is simply described as summation of interatomic bond energy parameterized by ab initio calculations, and the second term corresponds to the contribution of electrons in the surface dangling bonds [4, 5]. Atomic arrangements are generated according to the following rules: (i) Select atoms at random; (ii) select events as described in assumptions (3) and (4) at random; (iii) calculate the change in system energy ΔE after exchanging the chosen atoms; (iv) if ΔE is negative, accept the new configuration; (v) otherwise, select a random number R uniformly distributed over the interval (0,1); (vi) if $\exp(\Delta E/k_B T) < R$, accept the old configuration; (vii) otherwise, use the new configuration and the new system energy as the current properties of the system. This procedure is repeated for suitable number of configurations (MC steps) in order to approach equilibrium configurations at $T = 1225$ K. The appearance ratio of the structure in the AlN monolayer is obtained by 1000 samples at each III/V ratio in the range from 0.01 to 100.

Figure 7.13a displays the estimated appearance ratio of the Al–N dimer structure and the 4H–AlN obtained by the ECMC simulations as a function of the III/V ratio [26]. This implies that the larger the III/V ratio, more favorable the 4H–AlN. Most of the resultant atomic arrangements exhibit the 4H–AlN on the 4H–SiC($11\bar{2}0$) at the III/V ratio greater than 10. On the other hand, the Al–N dimer structure mainly appears at the III/V ratio less than 0.1, where the Al–N dimer structure different from the 4H–AlN appears in 51.3% of the simulated atomic arrangements at the III/V ratio of 0.01. This is because many N adatoms tend to form the stable Al–N dimer structures after Al adsorption as shown in Fig. 7.11b. The 4H–AlN still remains and the 2H–AlN appears at most in 6.5% as shown in Fig. 7.13b even under N-rich conditions, since the Al adatoms generally favor the 4H-sites and the N adatom occupying the F site for the 2H–AlN formations is energetically less

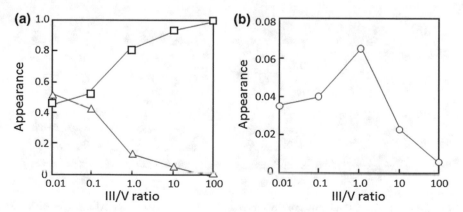

Fig. 7.13 Appearance ratio of **a** the 4H–AlN (open square) and the Al–N dimer structures (open triangle), and **b** the 2H–AlN (open circle) obtained by the ECMC simulations at 1225 K as a function of the III/V ratio

favorable by 0.18 eV than that occupying the G site for the 4H–AlN formations. Considering that the Al–N dimer structures are likely not to inherit the atomic arrangement of the 4H–SiC substrate, 55% of the AlN monolayer grown on the 4H–SiC(11$\bar{2}$0) may form the most stable 2H–AlN thin films at the III/V ratio of 0.01. Therefore, these results shown in Fig. 7.13 are qualitatively consistent with experimental results where the 4H–AlN and the 2H–AlN are grown under Al-rich and N-rich conditions, respectively [21].

The prototypical growth processes of the 4H–AlN at the III/V ratio of 100 and the 2H–AlN at the III/V ratio of 0.01 are shown in Fig. 7.14 [26]. The 4H–AlN formation proceeds according to Fig. 7.14a–d as follows. Under Al-rich condition, Al adatoms occupy the 4H-sites and the H sites having single bond with surface C atom as shown in Fig. 7.14a. This is because the Al adatom with smaller number of valence electrons favors the interatomic bonds with C surface atoms having two electrons in their dangling bonds by occupying the 4H-sites. Subsequent N adsorption forms Al–N pair with the 4H arrangements in Fig. 7.14b, c, since Al adatoms can easily migrate independently from N adsorption to occupy the 4H-sites as shown in Fig. 7.11a. Therefore, Al adatoms at the D site in Fig. 7.14b migrate only to the stable 4H-site (E site) in Fig. 7.14c. Consequently, the 4H–AlN appears under Al-rich condition. Under N-rich condition, however, N adatoms occupy the interstitial site located above top C atom, since the N adatom with larger number of valence electrons can form extra interatomic bonds with surface Si atoms without electrons in their dangling bonds. The N adatom located at the D site having two interatomic bonds with surface Si atoms can intermittently migrate to the 2H-site (F site) in Fig. 7.14g after Al adsorption. Thus some of N adatoms are rearranged to form Al–N pair with the 2H-arrangements in Fig. 7.14h. Although details of the AlN growth processes on the 4H–SiC(11$\bar{2}$0) substrate should be verified more quantitatively incorporating multi-layer growth in the ECMC simulations with larger unit cell, this is qualitatively consistent with the experimental fact that the

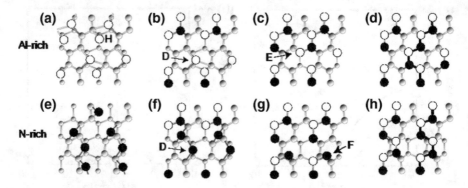

Fig. 7.14 Simulated results of growth processes of the 4H–AlN under Al-rich conditions in **a–d** and the 2H–AlN under N-rich conditions in **e–h**, where Al and N adatoms fully occupy the surface sites in advance as shown in **a** and **e**, respectively. Large black and white circles denote N and Al atoms, respectively

2H- and 4H–AlN films are grown under N- and Al-rich conditions, respectively [21, 22].

References

1. T. Zywietz, J. Neugebauer, M. Scheffler, Adatom diffusion at GaN(0001) and (0001) surfaces. Appl. Phys. Lett. **73**, 487 (1998)
2. L. Lymperakis, J. Neugebauer, Large anisotropic adatom kinetics on nonpolar GaN surfaces: consequences for surface morphologies and nanowire growth. Phys. Rev. B **79**, 241308 (2009)
3. A. Ishii, First-principles study for molecular beam epitaxial growth of GaN(0001). Appl. Surf. Sci. **216**, 447 (2003)
4. T. Ito, K. Shiraishi, A Monte Carlo simulation study on the structural change of the GaAs (001) surface during MBE growth. Surf. Sci. **357–358**, 486 (1996)
5. T. Ito, K. Shiraishi, Electron counting Monte Carlo simulation of the structural change of the GaAs(001)-c (4×4) surface during Ga predeposition. Jpn. J. Appl. Phys. **37**, 262 (1998)
6. T. Akiyama, D. Obara, K. Nakamura, T. Ito, Reconstructions on AlN polar surfaces under hydrogen rich conditions. Jpn. J. Appl. Phys. **51**, 018001 (2012)
7. T. Akiyama, Y. Saito, K. Nakamura, T. Ito, Reconstructions on AlN nonpolar surfaces in the presence of hydrogen. Jpn. J. Appl. Phys. **51**, 048002 (2012)
8. C.D. Lee, Y. Dong, R.M. Feenstra, J.E. Northrup, J. Neugebauer, Reconstructions of the AlN (0001) surface. Phys. Rev. B **68**, 205317 (2003)
9. J.E. Northrup, R. Di Felice, J. Neugebauer, Atomic structure and stability of AlN(0001) and (000$\bar{1}$) surfaces. Phys. Rev. B **55**, 13878 (1997)
10. M.S. Miao, A. Janotti, C.G. van de Walle, Reconstructions and origin of surface states on AlN polar and nonpolar surfaces. Phys. Rev. B **80**, 155319 (2009)
11. R. Miyagawa, S. Yang, H. Miyake, K. Hiramatsu, Effects of carrier gas ratio and growth temperature on MOVPE growth of AlN. Phys. Status Solidi C **9**, 499 (2012)
12. A. Rice, R. Collazo, J. Tweedie, R. Dalmau, S. Mita, J. Xie, Z. Sitar, Surface preparation and homoepitaxial deposition of AlN on (0001)-oriented AlN substrates by metalorganic chemical vapor deposition. J. Appl. Phys. **108**, 043510 (2010)
13. T. Akiyama, K. Nakamura, T. Ito, *Ab initio*-based study for adatom kinetics on AlN(0001) surfaces during metal-organic vapor-phase epitaxy growth. Appl. Phys. Lett. **100**, 251601 (2012)
14. J. Stellmach, M. Frentrup, F. Mehnke, M. Pristovsek, T. Wernicke, M. Kneissl, MOVPE growth of semipolar (11$\bar{2}$2) AlN on m-plane (10$\bar{1}$0) sapphire. J. Cryst. Growth **355**, 59 (2012)
15. Y. Takemoto, T. Akiyama, K. Nakamura, T. Ito, Ab initio-based study for surface reconstructions and adsorption behavior on semipolar AlN(11$\bar{2}$2) surfaces during metal–organic vapor-phase epitaxy growth. Jpn. J. Appl. Phys. **54**, 085502 (2015)
16. T. Akiyama, T. Yamashita, K. Nakamura, T. Ito, *Ab initio*-based study for adatom kinetics on semipolar GaN(11$\bar{2}$2) surfaces. Jpn. J. Appl. Phys. **48**, 120218 (2009)
17. T. Akiyama, Y. Takemoto, K. Nakamura, T. Ito, Theoretical investigations of initial growth processes on semipolar AlN(11$\bar{2}$2) surfaces under metal-organic vapor-phase epitaxy growth condition. Jpn. J. Appl. Phys. **55**, 05FA06 (2016)
18. S. Ichikawa, M. Funato, S. Nagata, Y. Kawakami, Optimal growth pressure for high quality semipolar AlN(1$\bar{1}$02) homoepitaxial films, in *The 60th JSAP Spring Meeting, 29a-G21-6* (2013)

19. Y. Takemoto, T. Akiyama, K. Nakamura, T. Ito, Systematic theoretical investigations on surface reconstruction and adatom kinetics on AlN semipolar surfaces. e-J Surf. Sci. Nanotech. **13**, 239 (2015)

20. N. Onojima, J. Suda, H. Matsunami, 4H-polytype AlN grown on 4H-SiC($11\bar{2}0$) substrate by polytype replication. Appl. Phys. Lett. **83**, 5208 (2003)

21. M. Horita, J. Suda, T. Kimoto, Impact of III/V ratio on polytype and crystalline quality of AlN grown on 4H-SiC($11\bar{2}0$) substrate by molecular-beam epitaxy. Phys. Status Solidi C **3**, 1503 (2006)

22. J. Suda, M. Horita, R. Armitage, T. Kimoto, A comparative study of nonpolar a-plane and m-plane AlN grown on 4H-SiC by plasma-assisted molecular-beam epitaxy. J. Cryst. Growth **301–302**, 4100 (2007)

23. T. Ito, T. Akiyama, K. Nakamura, T. Ito, "Systematic theoretical investigation for adsorption behavior of Al and N atomson 4H-SiC($11\bar{2}0$) surfaces. Appl. Surf. Sci. **256**, 1160 (2009)

24. M.D. Pashley, K.W. Haberern, W. Friday, J.M. Woodall, P.D. Kirchner, Structure of GaAs (001) (2×4)-c(2×8) determined by scanning tunneling microscopy. Phys. Rev. Lett. **60**, 2176 (1988)

25. T. Araki, T. Akiyama, K. Nakamura, T. Ito, Theoretical investigation of the structural stability of zinc blende GaN thin films. e-J Surf. Sci. Nanotech. **3**, 507 (2005)

26. T. Ito, T. Ito, T. Akiyama, K. Nakamura, *Ab initio*-based Monte Carlo simulation study for the structural stability of AlN grown on 4H-SiC($11\bar{2}0$). e-J. Surf. Sci. Nanotech. **8**, 52 (2010)

Chapter 8
Polarity Inversion and Electron Carrier Generation in III-Nitride Compounds

Takashi Nakayama

In this chapter, we consider two topics based on theoretical calculations; the surface polarity inversion during the film growth and the electron-carrier generation by structural defects in III-nitride compounds. Some of unique features of III-nitride compounds originate from their wurtzite crystal structure. Because of the lack of inversion symmetry along the c direction, III-nitride compounds become polar systems and have two kinds of surfaces; the (0001) III-group metal-polarity surface and the $(000\bar{1})$ N-polarity surface. By controlling the growth condition, the surface can be converted from one polarity surface to the other polarity. In Sect. 8.1, we adopt AlN as an example and consider the microscopic mechanism of the surface-polarity inversion during the AlN growth. Another fascinating feature of III-nitride semiconductor family is a wide range of their band-gap energies, 0.6–6.3 eV. Among III-nitrides, InN has the largest band width of the conduction band and thus the smallest band gap reflecting the large orbital radius of In atoms. Even when we produce InN films with special cares, there frequently appear unintentional electron carriers in InN films and around their interfaces/surfaces. In Sect. 8.2, we consider edge dislocations in InN films and show that such dislocations are one of origins for electron carrier generation. In Sect. 8.3, on the other hand, we consider Schottky barrier at metal/InN interfaces and show that electron carriers are also produced at these interfaces. Finally, we explain that the electron-carrier generation is closely related to the anomalous position of the charge neutrality level of InN.

T. Nakayama (✉)
Department of Physics, Chiba University, Chiba, Japan
e-mail: nakayama@physics.s.chiba-u.ac.jp

© Springer International Publishing AG, part of Springer Nature 2018   145
T. Matsuoka and Y. Kangawa (eds.), *Epitaxial Growth of III-Nitride Compounds*,
Springer Series in Materials Science 269,
https://doi.org/10.1007/978-3-319-76641-6_8

8.1   Surface Polarity Inversion

III-nitride compounds, AlN, GaN, and InN, have a wurtzite structure, different from most of semiconductors having a zincblende structure. This difference in crystal structure promotes unique physical properties in III-nitride systems [1]. With respect to the crystal growth, because of the lack of inversion symmetry along the c-axis direction, III-nitride compounds having wurtzite structure have two types of growing surfaces; the (0001) III-group metal-polarity surface and the ($000\bar{1}$) N-polarity surface. The metal-polarity surface is terminated by three-coordinated III-family metal atoms with one dangling bond directed to the +c direction, while the N-polarity surface is terminated by three-coordinated N atoms with one dangling bond directed along +c. These surfaces exhibit different surface morphology, growth kinetics, and optical/conductive and chemical properties [2]. In the case of GaN, for example, the Ga-polarity surface is usually very smooth, thus being preferred to produce films with high uniformity and crystallinity. On the other hand, the N-polarity surface often shows rough and hexagonal pyramid morphologies but have high chemical reactivities; the former feature is useful to fabricate nanowires and nanodots, [3, 4] while the latter feature enables the higher impurity-atom incorporation such as In atoms to produce InGaN [5, 6]. Recent advances of polarity control technologies succeeded in the artificial production of periodic polarization domains useful for nonlinear optics [7]. In this way, the control of surface polarity is essential to produce high-quality films and desirable device structures.

One of interesting phenomena with respect to the polar surfaces is the surface polarity inversion during the film growth. In the metal organic vapor phase epitaxy (MOVPE) process, Lim et al. observed that the GaN films exhibited the N-polarity surfaces when GaN were grown on the N-polarity AlN substrate, which was obtained by H_2-cleaning and nitriding the sapphire substrate [8]. On the other hand, they found that, when a few monolayer Al atoms were intentionally supplied on such N-polarity AlN substrate, the surface of GaN films was converted from the N polarity into the Ga polarity. Figure 8.1a shows the observed spectra by a coaxial impact collision ion scattering spectroscopy (CAICISS), [2] where we can see that the grown surface changes from the N polarity (c-) to the Ga polarity (c+) as increasing the supplying time of Al atoms on the substrate. The similar inversion was also observed in the GaN growth on (Al, Ga, In)-pretreated GaN [9].

In the case of III-nitride semiconductor growth on metal-supplied N-polarity substrates, considering the pretreatment stacking time of metal atoms, Lim et al. proposed that the polarity inversion occurs by the appearance of interface metal-atom layers with two-monolayer thickness as shown in Fig. 8.1b [8]. The validity of this two-monolayer model was also confirmed by optical experiments [9]. In this section, we selected AlN surfaces and pretreated Al metal overlayers, and clarified the mechanism of the surface-polarity inversion, by using the first-principles theoretical calculations [10].

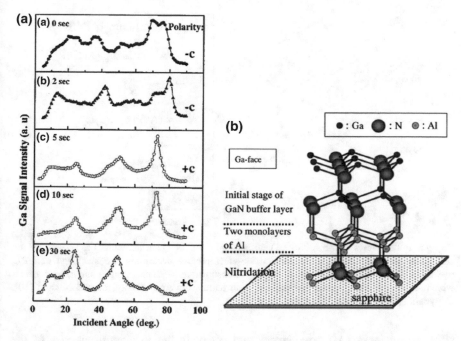

Fig. 8.1 a Co-axial impact collision ion scattering spectroscopy (CAICISS) spectra of GaN film surface after different presupply times of Al atoms as a function of the incident angle of the He$^+$ ion beam. We can clearly observe that, with increasing presupply time, the surface changes from the N polarity (-c) to the Ga polarity (+c). **b** Al-two-monolayer model to realize the surface polarity inversion. Reproduced with permission from Lim et al. [8]. Copyright (2002) by American Institute Physics

8.1.1 Modeling of Inversion

The stable reconstructed surface structure is known to be the (2×2) for AlN [11]. However, since no superstructures were observed for the surfaces during the AlN growth, we assume to adopt the (1×1) surface unit shown in Fig. 8.2a for simplicity, and employ the standard repeated-slab geometry to simulate the surfaces during the AlN growth. In this modeling, all the atoms in the substrate and adsorbed layers are located at T1, T4, or H3 site in this figure owing to the symmetry. Two-bilayer (0001) AlN slab was used as the N-polarity AlN substrate, where the back Al-polarity surfaces are terminated with virtual hydrogens to eliminate the influence from the back surface [12].

We first stack $(Al)_m$ overlayers on this N-polarity substrate as shown in Fig. 8.2b, and calculate the formation energies of the $(Al)_m/(AlN\text{-substrate})$ systems to examine which structure is more stable in $(Al)_m$ overlayers. This is because Al metal layers are usually much softer than the AlN substrate and Al atoms easily diffuse on the surface, thus the most stable structure is obtained by the step-by-step stacking of Al atoms. Then, we stack $(AlN)_1$ and $(AlN)_2$ layers of the wurtzite

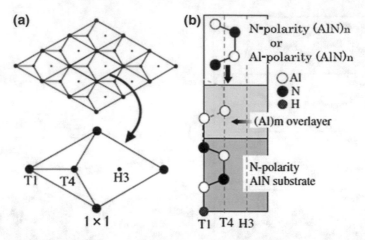

Fig. 8.2 **a** Top views of the AlN(0001) surface and the (1 × 1) surface unit. T1, T4, and H3 are adsorption sites of atoms. **b** Schematic side-view picture describing the stackings of (Al)$_m$ overlayers and (AlN)$_n$ layers on the N-polarity AlN substrate. With changing m and n, we consider which stacking is more stable. Reproduced with permission from Nakayama and Mikami [10]. Copyright (2005) by John Wiley and Sons

structure on these (Al)$_m$ overlayers, and calculate the formation energies of the (AlN)$_n$/(Al)$_m$/(AlN-substrate) systems to examine which polarity surface is more stable in the top (AlN)$_n$ layers.

In order to determine the stable atom positions and to calculate the formation energies of the (Al)$_m$/(AlN-substrate) and (AlN)$_n$/(Al)$_m$/(AlN-substrate) systems, the total-energy pseudopotential method is employed in the local density approximation using the TAPP code, [13] where all atom positions are optimized. This method is a standard one explained in other chapters and the details are also described elsewhere [10, 14].

8.1.2 Stability of Al Overlayers

First, we consider the stacking of (Al)$_m$ overlayers on the N-polarity AlN substrate. Figure 8.3a shows the formation energies of (Al)$_m$/(AlN-substrate) systems, for various adsorption positions of Al atoms and Al-overlayer thicknesses, m. It is seen that when one- and two-monolayer amount of Al atoms are supplied on the AlN substrate adsorbed Al atoms prefer to be located first at T1 site and then at T4 or H3 site, as shown by a solid-line sequence. These sites correspond to atom positions of the wurtzite or zincblende structure having the four-folded coordination for Al atoms. This result indicates that the atom position of (Al)$_m$ overlayers copies the AlN-substrate wurtzite structure such that the (Al)$_2$ unit imitates the ionic unit of AlN.

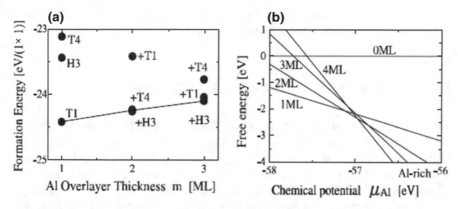

Fig. 8.3 **a** Calculated formation energies of (Al)$_m$ overlayers on the N-polarity AlN substrate as a function of m. The symbols denote the adsorption sites of Al atoms. Second and third layer Al atoms are stacked on the most stable (Al)$_{m-1}$ layers, thus the T1-T4-H3 stacking is the most stable. **b** Calculated free energies of the most stable m-monolayer (Al)$_m$ on the N-polarity AlN substrate, which is denoted as mML, as a function of a chemical potential of supplied Al atoms. Reproduced with permission from Nakayama and Mikami [10]. Copyright (2005) by John Wiley and Sons

On the other hand, when three-monolayer amount of Al atoms are supplied, the Al atom at the top surface prefers to be located at H3 or T1 site, which corresponds to the atom position of Al fcc-bulk system. This stacking occurs because the top-surface Al atom is bound to underlying Al atoms with the metallic bondings, thus being released from the crystal-structure heredity of AlN substrate. In this way, (Al)$_m$ overlayers follow the substrate wurtzite structure till two-monolayer thickness, $m \leq 2$, while they produce the fcc-bulk structure in thicker cases of $m \geq 3$. This kind of change of layer feature was also examined for GaN surface, where the Ga overlayers show liquid-like softening along the surface when their thickness is larger than two [15].

Next, we consider whether the Al overlayer with two-monolayer thickness is easy to be realized in experiments or not. Figure 8.3b shows the calculated free energies of (Al)$_m$ overlayers on N-polarity AlN substrate as a function of a chemical potential of supplied Al atoms. The basic idea of this analysis is described in literatures [14, 16]. It is seen that as the chemical potential increases corresponding to the Al-richer condition, the surface changes from the clean AlN surface to the surfaces with adsorbed Al atoms of one, two, and then four monolayers (not three), and that the Al overlayer with two-monolayer thickness definitely has a finite stable region with respect to the Al-atom supply. However, the calculated chemical potential of fcc-bulk Al is −57.2 eV, which roughly corresponds to the accumulation onset of bulk Al phase on the surface and is located at the left edge of the stable region of Al overlayers with two-monolayer thickness. These results indicate that the surface with two-monolayer Al atoms is unstable and is difficult to be realized in thermodynamic equilibrium. In other words, as pointed out in [8], one has to use the growth process far from thermodynamic equilibrium to realize the surface with two-monolayer Al atoms.

8.1.3 Stability of Polarity-Converted AlN Layers

Then we consider the stacking of $(AlN)_n$ layers on the $(Al)_m$ overlayers that are stacked on the N-polarity AlN substrate. The formation energies of $(AlN)_n/(Al)_m/$ (AlN-substrate) systems are shown in Fig. 8.4a, as a function of $(Al)_m$-overlayer thickness, m. It is seen that the stacked $(AlN)_n$ layers have the stable converted Al-polarity surface for $m = 2$, while the N-polarity surfaces are still stable for $m = 1$ and 3. To understand the reason of this variation, we consider the stacking of $(AlN)_2$ layers on the hypothetical $(Al)_m$ substrate having either fcc-bulk or wurtzite structure with the same lattice constant as AlN along the interface. Figure 8.4b shows the calculated formation energies of these $(AlN)_2/(Al)_m$ systems, as a function of the substrate thickness, m. In the case of fcc-bulk $(Al)_m$ substrate, the N-polarity surface is always stable independent of the substrate thickness, m. On the other hand, in the case of wurtzite $(Al)_m$ substrate, the N-polarity surface is stable for the odd number of m, while the Al-polarity surface is stable for $m =$ even-number.

The difference of stacking-energy variation seen in Fig. 8.4b is explained by considering the stability of $(AlN)_2/(Al)_m$ interfaces and the transcription effect of the underlying $(Al)_m$-substrate structure. First of all, since the binding energy is larger for Al–N bonds than for Al–Al ones, in all the cases, the $(AlN)_n$ layers are first adsorbed on the underlying $(Al)m$ substrates such that the N atom in $(AlN)_2$ connects to the substrate-top Al atom in $(Al)_m$ to stabilize the $(AlN)_2/(Al)_m$ interface.

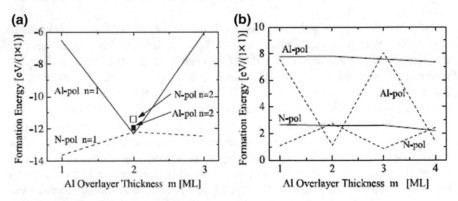

Fig. 8.4 Calculated formation energies of **a** $(AlN)_n/(Al)_m/(AlN$-substrate) and **b** $(AlN)_2/(Al)_m$ systems as a function of m. Al-pol and N-pol indicate the surface polarities of stacked $(AlN)_n$ layers. Solid and dashed lines in **b** correspond to the fcc-bulk and wurtzite $(Al)_m$ substrates, respectively. Reproduced with permission from Nakayama Mikami [10]. Copyright (2005) by John Wiley and Sons

In the case of fcc-bulk $(Al)_m$ substrate, reflecting the fcc-bulk structure, such N atom in $(AlN)_2$ is located at the oblique position compared to that of substrate-top Al atom, as shown in Fig. 8.5a, b. Since the interface geometry is the same for both $m =$ odd-number and $m =$ even-number cases, the staked $(AlN)_2$ layers always have the N-polarity surface not depending on the thickness of underlying $(Al)_m$ substrate. On the other hand, in the case of wurtzite $(Al)_m$ substrate, the N atom in $(AlN)_2$ is located just above the substrate-top Al atom as seen in Fig. 8.5c for the even number of m, while such N atom takes an oblique position as in Fig. 8.5d for the odd number of m. Thus, the stacked $(AlN)_2$ layers have the Al- and N-polarity surfaces for $m =$ even-number and $m =$ odd-number, respectively. From these considerations, we can say that the difference of $(AlN)_n$-stacking variation between fcc-bulk and wurtzite $(Al)_m$ substrates originates from the difference of crystal-structure periodicity along the (0001) direction between two kinds of substrates; the fcc-bulk structure has a monolayer period, while the period of wurtzite structure is two monolayers.

From these results, one can conclude that, only when the Al overlayer has two-monolayer thickness and such Al overlayers forms the wurtzite structure, the surface-polarity inversion occurs during the following AlN growth. This conclusion supports the appropriateness of the Al-two-monolayer model proposed in [8] to realize the surface-polarity inversion. However, Al overlayers with more than two monolayers cause no surface-polarity inversion because they form the fcc bulk structure. Recently, Harumoto et al. deposited AlN on Pt and Al (111) fcc metal substrates and found that grown AlN layers respectively show the N- and Al-polarity surfaces [17]. One of reasons for the discrepancy between this experiment and present theoretical results in Fig. 8.5a, b might be related to the strain effect of the substrate on the film growth. In the present calculation, the hypothetical fcc-bulk and wurtzite $(Al)_m$ substrates are coherently strained so as to match the lattice constant to underlying AlN substrate. In fact, it is well known that the strain

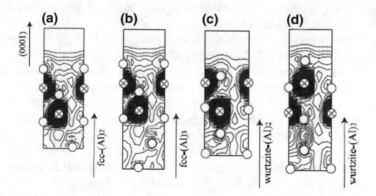

Fig. 8.5 Calculated stable stacking geometries of $(AlN)_2$ layers on $(Al)_m$ substrates. **a** and **b**: fcc-bulk $(Al)_m$ substrate with $m = 2$ and 3, **c** and **d**: wurtzite $(Al)_m$ substrate with $m = 2$ and 3, respectively. Open and crossed circles denote Al and N atoms, respectively. Reproduced with permission from Nakayama Mikami [10]. Copyright (2005) by John Wiley and Sons

affects the growth processes and changes grown films such as having different stoichiometry [18].

It should be noted here that, in the case of Al overlayer with two-monolayer thickness, the Al atom in the upper layer can be located either at T4 or H3 site, as shown in Fig. 8.3a, which sites respectively corresponds to the wurtzite or zinc-blende structure. It is quite natural for Al overlayers to have either wurtzite or zincblende structure because the formation energy difference between both structures is generally small in cases of semiconductors with small iconicity [1]. This result indicates that, when the surface-polarity inversion is realized, the stacking fault is expected to appear between the upper and lower different polarity AlN layers with a possibility around 50%.

8.1.4 Concluding Remarks

In this section, we concentrate on the polarity inversion in the case of III-nitride compounds. However, the similar inversion was also observed in other semiconductors such as ZnO on GaN [19] and AlN on oxidized AlN [20]. In these systems, the oxide layers like Ga_2O_3 and Al_2O_3 with inversion symmetry are expected to work for the polarity inversion. These interface layers are stable because they are semiconductors, which is different from the present metallic Al overlayers.

Then, we discuss electronic structures of $AlN/(Al)_2/AlN$ sandwich systems. Al atoms at Al/AlN interfaces have the sp^3 electron configuration. Thus the interface Al atoms in $(Al)_2$ overlayers produce covalent bonds with N atoms in lower and upper AlN layers and show semiconducting properties perpendicular to the interface, while such interface Al atoms in $(Al)_2$ overlayers produce metallic bonds with other Al atoms in the same overlayers along the interface and thus are expected to show properties of two-dimensional metals. On the other hand, since the wurtzite AlN layers are symmetrically arranged on both sides of two-monolayer $(Al)_2$ overlayers, there is little band offset between lower and upper AlN layers. However, since the Fermi's energy is located within the band gap of AlN, the $(Al)_2$ overlayers work as potential wells for both electrons and holes in AlN.

Finally, we shortly comment on the realization of other interesting converted interfaces. As explained in the above, the polarity inversion of grown films occurs depending on the thickness and strain of intermediate layers. By controlling these factors during the film growth, we can expect the conversion of wurtzite substrate into zincblende films, which is also realized using the boundary effects in the case of III-nitride nanowires. At wurtzite/zinc blende interfaces, we can produce the definite band offset by using the surface polarization existing only in wurtzite layers and by using the difference in bond network topology between wurtzite and zinc blende layers [1]. Such interfaces are expected useful for designing hetero devices realizing strong carrier confinement with little structural defects.

8.2 Electron Carrier Generation by Dislocation

III-nitride compounds and their alloys are promising constituent materials for optoelectronic devices because their band gaps cover a wide range of wavelength from the ultraviolet (~ 200 nm) to infrared (~ 2 mm) regions. Among these materials, due to the small band gap of about 0.6 eV, InN plays an important role in the infrared regions [21, 22]. InN is often grown by molecular beam epitaxy (MBE) and MOVPE. Owing to the developments of crystal growth techniques, the crystal quality of grown InN films has made remarkable progress such as having large-area atomic-order smooth and flat surfaces. However, it has long been observed that there always appear unintentional electron carriers even when one produces InN films with special cares. These carriers seem to occupy about 0.1–0.9 eV of the conduction-band bottom, [22, 23] thus preventing the realization of p-type InN.

Unintentional carriers are considered to be created by point defects such as atom vacancies because the lattice mismatch around 10% always exists between the substrate and the grown InN layers in the present stage of crystal growth. Wang et al. found by performing the Hall measurement together with counting the number of dislocations in samples that the edge-type threading dislocations also produce a number of electron carriers, with the density over 10^{17} cm^{-3} [24]. Figure 8.6 shows the electron concentration and Hall mobility of a number of InN films as a function of full width at half maximum (FWHM) values of x-ray rocking curves. Here, the FWHM represents the density of dislocations. It is clearly seen that, with increasing the dislocation density, the electron concentration increases and the Hall mobility decreases due to the scattering by such dislocations.

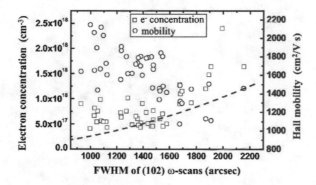

Fig. 8.6 Electron concentration and Hall mobility of a number of InN films as a function of full width at half maximum (FWHM) values of x-ray rocking curves. Dashed line denotes the estimated density of dangling bonds originated from edge-component threading dislocations. It is clearly seen that, with increasing FWHM, i.e., the density of dislocations, electron concentration increases and Hall mobility decreases. Reproduced with permission from Wang et al. [24]. Copyright (2007) by American Institute Physics

Since dislocations often work as deep-level electron/hole traps and never produce movable carriers in conduction and valence bands of most semiconductors, [25] it is interesting to study whether dislocations can become donors in the case of InN films as suggested by experiments. In this section, we study electronic structures of InN edge dislocations and consider the origin of carrier generation by dislocations, based on the results using the first-principles theoretical calculations, comparing with the cases of other point defects like vacancy and other semiconductors like GaN [26].

8.2.1 Modeling of Dislocation

The atomic structures of edge dislocations in III-nitride compounds were studied by Béré and A. Serra, and Lei et al., using the empirical atomistic elastic calculations [27, 28]. Figure 8.7a–c schematically show core structures of such dislocations. Since these have 5 + 7-, 8-, and 4-folded atomic rings perpendicular to the c-axis, respectively, contrary to the six-folded ring networks seen in bulk InN, we hereafter call these the 5/7 core, 8 core, and 4 core dislocations.

These dislocations have the same periodicity along the c-axis, i.e., the dislocation direction, as that of the hexagonal bulk system. However, they have no special periodicity perpendicular to the c-axis, thus it is difficult to adopt the conventional

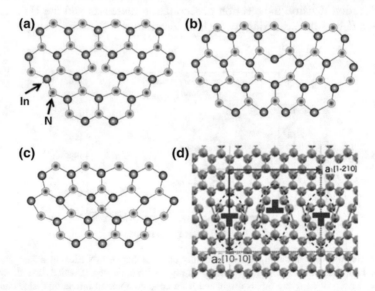

Fig. 8.7 Schematic pictures of dislocations in InN, viewed from the c- direction. **a** 5/7 core, **b** 8 core, and **c** 4 core dislocations. **d** Repeated unit cell used in the first-principles calculation, for the case of 5/7 core dislocation. Reproduced with permission from Takei and Nakayama [26]. Copyright (2009) by Elsevier

electronic-structure calculations using the periodic unit cells. There are two typical methods to treat the dislocation; one is to use the cluster geometry, [25] where the dislocation core is embedded in a bulk pillar and the pillar surface is terminated by virtual hydrogen atoms. The other method is to employ the repeated unit-cell systems that have a pair of dislocations with opposite directions in one-unit cell as shown in Fig. 8.7d, which is the case of 5/7 core dislocation [26]. With increasing the system size, both methods give the similar results. Hereafter, we show the results using the latter method.

Atomic and electronic structures of dislocations in InN are calculated by the standard first-principles total-energy method in the generalized gradient approximation (GGA), using Vanderbilt-type ultrasoft pseudopotentials for both In (5s, 5p, 5d) and N (2s, 2p) atoms, [29] the PBE form exchange-correlation potential, [30] 25 ryd cutoff energy for plane-wave expansion of wavefunctions, and the TAPP code [13]. All atomic positions and unit-cell size are optimized.

Here, we shortly comment on the relative stability of dislocations. The calculated formation energies are 1.92, 2.30, and 2.23 eV/Å for 5/7 core, 8 core, and 4 core dislocations, respectively, when the present repeated unit-cell model is used. Of course, this calculation does not consider the elastic energy loss apart from the core region. However, because the difference of such energy is normally small between dislocations compared to the energy around the core, [28] we can expect that the 5/7 core dislocation has the lowest formation energy. In real films, however, since the formation energy strongly depends on the strain environment and growth process and it is difficult to decide which dislocation structure is more stable, we hereafter consider all electronic structures of the above-mentioned three dislocations.

8.2.2 Electronic Structures and Carrier Generation

We first show the calculated band structure of bulk InN in Fig. 8.8a, which are obtained using the same large unit cell as the cases of dislocation to make a comparison easier. The bands, c1 and v1 in this figure, are the lowest conduction and highest valence bands of bulk InN, respectively. The band-gap energy here is 0.25 eV, which is smaller than the experimental results of about 0.6 eV [31]. This occurs because the band gap is often underestimated by the calculations based on the density-functional theory (DFT) [32]. It should be noted here that the conduction band has a markedly large band width reflecting the large atomic radius of In atom. This feature is the most important key to understand various electronic properties of InN, as shown in the followings.

Then, we consider the 5/7 core dislocation, whose calculated band structure is shown in Fig. 8.8b. The c1 and v1 bands touch with each other around the Γ point, which indicates some shrinkage of the band-gap energy around the dislocation. Most remarkable feature in this figure is the position of the Fermi energy, E_F at 0 eV; E_F is located about 0.6 eV above the bottom of the lowest conduction band

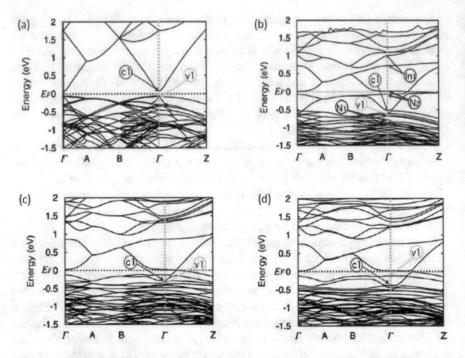

Fig. 8.8 Calculated band structures of **a** bulk InN and dislocations in InN with **b** 5/7 core, **c** 8 core, and **d** 4 core. The same unit cell is used to make the comparison easy. c1 and v1 are the lowest conduction band and the highest valence band of bulk InN, respectively. N1 and N2 are bonding and anti-bonding states of N dimer bonds, while In1 denotes the bonding state of In dimer bonds. E_F at 0 eV denotes the Fermi-energy position. Reproduced with permission from Takei and Nakayama [26]. Copyright (2009) by Elsevier

(c1) of bulk InN. This result indicates that the electron carriers are generated in the conduction band, in agreement with the experiment [24, 33].

In order to understand the origins of such electron carriers, we analyze all band states around the band gap. Figure 8.9a–d show the charge–density plots of electronic states at Γ, which are respectively denoted by c1, N1, N2, and In1 in Fig. 8.8b. It is seen that the c1 state is a conduction-band state of bulk InN, thus being an anti-bonding state of In-s and N-s orbitals and extended over the system. The N1 and N2 states are, respectively, bonding and anti-bonding states of dangling bonds of N atoms that are located in the center of dislocation. On the other hand, the In1 state is the bonding state of dangling bonds of In atoms in the core region. Since these states are strongly localized around the dislocation core, their bands show flat dispersions.

From these analyses, we obtain the overall feature of electronic structure of InN dislocation, which is schematically displayed in Fig. 8.10a. In the dislocation core, there exist three-folded In and N atoms. Since these In and N atoms have dangling bonds and are nearest to the same kinds of atoms, they produce N–N and In–In dimer bonds to stabilize the dislocation [25]. As seen in Fig. 8.8b, the bonding state

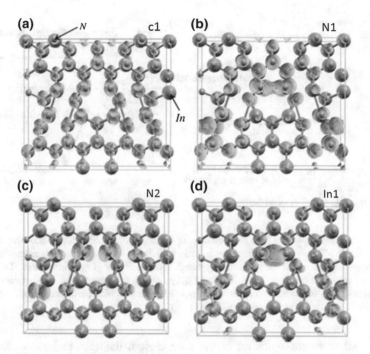

Fig. 8.9 Charge densities of electronic states at Γ of **a** c1, **b** N1, **c** N2, and **d** In1 bands shown in Fig. 8.8(b). Small and large balls indicate N and In atoms, respectively, while charge densities are described as green clouds. Reproduced with permission from Takei and Nakayama [26]. Copyright (2009) by Elsevier

of N–N dimer, N1, is located about 0.3 eV below the valence-band top, while the anti-bonding one, N2, appears about 0.5 eV above the conduction-band bottom. On the other hand, the bonding state of In–In dangling-bond dimer, In1, is located much higher, at least 1 eV, above the conduction-band bottom. Since the dangling bonds of N and In atoms have 1.25 and 0.75 electrons, the N2 and In1 bands are originally occupied by 0.5 and 1.5 electrons, respectively. Reflecting the relative energy positions of these bands, the In1 band becomes donor band and supplies the electron carriers to the conduction band of bulk InN and the N2 N–N anti-bonding band as shown by the down arrow. It should be noted here that the latter band is still occupied partially after such electron supply, which is contrary to the case of GaN shown below. As a result, the Fermi energy tends to be pinned at the energy position around the N2 band.

Then, we consider other types of dislocations. Figure 8.8c, d show the calculated band structures of 8 core and 4 core dislocations, respectively. Since the core of 8-ring dislocation also includes In and N dangling bonds, similar to the case of 5/7 core dislocation, the In dangling bonds supply the electron carriers to the conduction band similar to that shown in Fig. 8.10a and the Fermi energy is located about 0.2 eV above the conduction-band bottom. On the other hand, there are no In and N dangling bonds in the core of 4 core dislocation. However, due to the

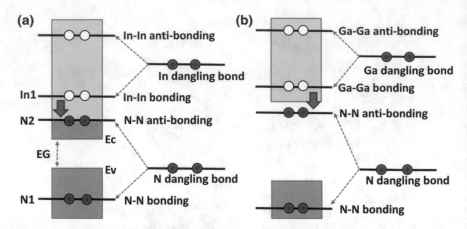

Fig. 8.10 Schematic pictures to explain electronic structures of 5/7 core dislocations in **a** InN and **b** GaN films. The vertical blue down-arrows indicate the electron transfer from the In-dangling-bond states to the conduction band and the N-dangling-bond states. Dark grey regions are occupied by electrons, while light grey regions are unoccupied. E_G between square boxes corresponds to the fundamental band gap. Reproduced with permission from Takei and Nakayama [26]. Copyright (2009) by Elsevier

compressed atom configuration in the core region, the valence-band states of N atoms localized around the core increase the energy to about 0.25 eV above the conduction-band bottom as seen in Fig. 8.8d and become the source of electron supply to the conduction band such as semi-metal materials. In this way, though the origin of electron-carrier generation is different between 8 core and 4 core dislocations, all these dislocations also become the source of electron generation.

8.2.3 Comparison with Other Defects

It is interesting to consider whether the same scenario of carrier generation applies to other defects in InN and dislocations in other semiconductors. We first consider

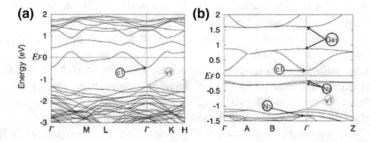

Fig. 8.11 Calculated band structures of **a** N vacancy in InN and **b** 5/7 core dislocation in GaN. Reproduced with permission from Takei and Nakayama [26]. Copyright (2009) by Elsevier

the point defect, i.e., N vacancy, in InN. Figure 8.11a shows the band structures of N vacancy, which was calculated using a bulk-InN unit cell with a single vacancy. Since the unit cell is somewhat small, $2 \times 2 \times 2$, the deformation of conduction band is large from the dispersion of bulk bands. However, it is seen that the Fermi energy is located in the conduction band, thus the electron carriers appearing. This is because there are four In dangling bonds around the N vacancy and they supply electrons to the conduction band similar to that shown in Fig. 8.10a. In this sense, we can say that the generation origin of electron carriers is the same between dislocations and N vacancy.

Then, we consider the dislocation in GaN. Figure 8.11b shows the calculated band structure of 5/7 core dislocation in GaN. As seen in this figure, the Fermi energy $E_F = 0$ is located in the fundamental band gap of GaN (between c1 and v1 bands) and there are no electron carriers in the c1 conduction band. This occurs because, as seen in the left of Fig. 8.10b, GaN has a large band gap around 3.5 eV and the flat bands made of N–N anti-bonding states are located deep in the fundamental band gap, thus, the Ga dangling bonds supply electrons not to the conduction band but only to the N–N anti-bonding states and the latter states become fully occupied. Thus, the dislocation core of GaN becomes only the acceptor and the electron carriers never appear in the conduction bands of GaN, being in good agreement with the experiments [27]. The similar trend to GaN is also seen for dislocations in cubic semiconductors such as Si and GaAs [25]. In this way, reflecting the large dispersion of the InN conduction bands, InN shows quite different electronic structure of defect states from those of GaN and AlN.

8.2.4 Nature and Inactivation of Electron Carriers

In this subsection, we first consider how many electrons are supplied from the dislocation core into the conduction bands of InN using the simplest model [34]. We employ the model because the above-mentioned first-principles calculations were performed using a small unit cell and thus do not apply to the real system having extended electron carriers. We first assume that the dislocation is a line positively charged with the line charge density, $\lambda = 0.17e/\text{Å}$, as shown in Fig. 8.12. In this case, the Hamiltonian for an electron around the dislocation is described as $\hat{H} = p^2/2m^* - \lambda/2\pi\varepsilon \cdot \log(r)$, where p is the momentum, r the radial coordinate, $m^* = 0.14$ the effective mass of electron, and $\varepsilon = 14.6$ the dielectric constant. The second term indicates the attractive Coulomb potential around the dislocation. Thus, using the uncertainty principle, the localization length of bound electron around the dislocation is estimated as $r_b \sim \sqrt{\hbar^2/m^* \cdot 2\pi\varepsilon/\lambda} \sim 14$ Å. On the other hand, there appear many electrons from the dislocation core and if these electrons are distributed uniformly in a cylinder with a $2r_b$ radius around the dislocation, the Tomas-Fermi screening length of potential becomes about $l \sim 7$Å, which is shorter than the localization length derived above. This estimation indicates that most of

Fig. 8.12 Schematic picture
of wavefunction $\psi(\mathbf{r})$ of
bound electron and potential
$V(\mathbf{r})$ around positively
charged dislocation in InN. r_b
and l are the localization
length of bound state and the
screening length of potential
by electron charges,
respectively

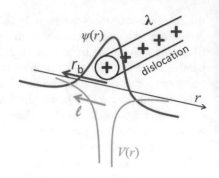

electrons produced from the dislocation core become movable free carriers in the conduction bands of InN.

It is interesting to note that the electron carriers coexist with charged dislocations. Thus, we can expect the unique transport properties in InN samples having dislocations [35]. In fact, Miller et al. demonstrated that the observed weak temperature dependence of the electron mobility and Seebeck coefficient is well explained by considering the scattering of electrons by charged dislocations [36]. In addition, Baghani and O'Leary argued that the non-uniform distribution of electron free carriers around dislocations is important to explain the observed values of electron mobility [37].

Next, we consider the realization of p-type InN. As shown above, the point defects and dislocations in InN produce unintentional electron carriers, which make it difficult to produce p-type InN for device applications. Considering that the electron density is around 10^{17} cm^{-3} for good crystalline InN films, one of methods to realize p-type InN is to dope acceptor atoms with more than this density. In fact, it was confirmed by Hall mobility, thermopower, electrolyte-based capacitance-voltage, and optical measurements that Mg-acceptor doping more than 10^{18} cm^{-3} density realizes the p-type conduction [33, 38]. However, the Mg-doping more than 5×10^{19} cm^{-3} produces complexes made of Mg and defects and again results in the n-type InN films.

Another method to realize p-type InN is to decrease electron carriers from the dislocation by doping C and O atoms into the dislocation core [34]. Figure 8.13a, b show the schematic energy diagrams of C- and O-doped InN 5/7 dislocations, respectively, which are obtained from calculated band structures by the first-principles method. In the case of C-atom doping, The In-C bonding states shown in Fig. 8.13c appear between the V1 and C1 states, i.e., in the band gap of InN, thus there is no electron supply to the conduction bands of InN. In the case of O-atom doping, the dangling-bond states of O atom shown in Fig. 8.13d appear in the band gap, thus there is also no electron supply to the conduction bands of InN. In this way, the C- and O-atom doping into the dislocation core eliminates the electron carriers caused by In dangling bonds. It has not been clear how to realize this doping and there have been no experimental trials responding to this theoretical prediction yet. However, since the elimination of unintentional electron carriers is

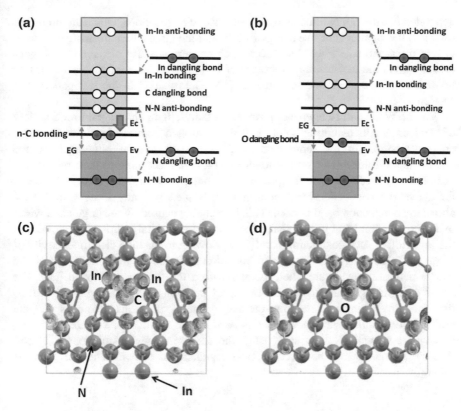

Fig. 8.13 Schematic energy diagrams of **a** C doped dislocation and **b** O doped dislocation in InN. Dark grey regions and filled circles indicate that their energy states are occupied by electrons, while light grey region and empty circles indicate that electron occupation is empty. The vertical blue down-arrow in **a** indicates the electron transfer. Charge densities (green clouds) of **c** electron occupied In-C bonding state and **d** electron occupied O dangling-bond state around dislocation in InN. Small and large balls indicate N and In atoms, respectively. Doped C and O atoms are located at the center of dislocation core

still an important issue to be solved, we need the study to decrease electron carriers based on the fundamental understanding of electronic structures of InN.

8.3   Schottky Barrier at Metal/InN Interface

Because of small band gap (about 0.7 eV) and small electron effective mass (0.14 m_0), InN is promising for optoelectronic device usage. To employ InN in practical devices, it is essential to produce metal/InN interfaces for carrier injection and to characterize the Schottky barriers at these interfaces. On the other hand, as explained in the previous section, unintentional electron carriers are easily generated in grown InN films by point defects and dislocations, which are produced owing to the smaller cohesive energy of InN and the large lattice mismatch between grown films and substrates. Moreover, such carriers are also observed to be generated around the surfaces and interfaces of InN [23, 38]. These features are closely related to unique electronic structure of InN as explained in the last part of this section.

Van de Walle and Neugebauer proposed a band-alignment diagram for a variety of semiconductors assuming boldly that the hydrogen impurity level has the same energy position in all semiconductors, [39] such as the alignment of the charge-neutrality levels of semiconductors [14, 40]. According to their results, the Fermi energy levels of most of representative metals, such as Al and Au, are located far above the conduction-band bottom of InN, i.e., typically more than 1.0 eV above the conduction band bottom. This is probably caused by the large band width of the conduction band in InN that reflects the large orbital radius of In atoms and the lowering of such conduction-band energy owing to the high electronegativity of N atoms. However, for most of metal/semiconductor interfaces, since the charge-neutrality levels are often located within the fundamental band gaps, the Fermi energies of metals match certain energy positions in the band gaps [14]. It is, therefore, interesting to determine the Fermi energy positions of metals that are realized at metal/InN interfaces. In this section, we consider the Schottky barrier behavior at metal/InN interfaces using the first-principles calculations [41] and discuss the relation to the electron carrier generation explained in the previous section.

8.3.1 Modeling of Interface

To simulate metal/InN interfaces, the standard repeated-slab geometry with a (1×1) surface unit cell is adopted for simplicity. We first prepared InN(0001) slab systems made of six InN layers, $(InN)_6$, with the In- and N-polarity surfaces. Then, seven layers of metal atoms are stacked at stable positions on these surfaces, as shown in Fig. 8.14a, b. Here, we chose Al and Au as representative metals because they have small and large work functions, respectively [42]. The In and N atoms on the back surfaces are terminated by hypothetical hydrogen atoms with fractional charges of 1.25 and $0:75e$ [12]. Therefore, we obtain slabs of $(metal)_7/(InN)_6H$. A vacuum region equivalent to the thickness of five monolayers is inserted between these slabs to isolate the interfaces.

Note that the stacked metal layers are pseudomorphic, i.e., coherently strained, owing to the lattice mismatch between bulk metals and InN, although they prefer to locate at the wurtzite sites till two-atomic-layer thickness and prefer to form fcc-bulk-like structures above the second layer as explained in Sect. 8.1 [10]. In fact, with increasing metal-layer thickness, we can reasonably expect that metal layers will gradually recover the bulk structure. However, since both bulk and pseudomorphic metal layers show metallic electronic structures, their Fermi

Fig. 8.14 Schematic atomic
structures at (1×1) Al/InN
interfaces viewed from the
$(11\bar{2}0)$ direction: **a** N-polarity
and **b** In-polarity interfaces.
Seven metal-atomic layers are
stacked at stable positions on
a six-layered InN substrate.
The back surface of InN is
terminated by hypothetical
hydrogen atoms. Reproduced
with permission from Takei
and Nakayama [41].
Copyright (2009) by the
Japan Society of Applied
Physics

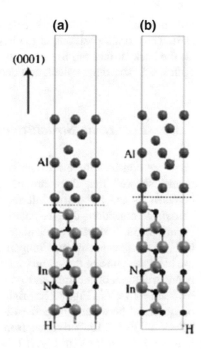

energies coincide with each other and the screening of electric fields occurs around
their interfaces. Thus, such stacking in the calculation is expected to produce
acceptable values for the Schottky barrier height (SBH) at real metal/semiconductor
interfaces [14, 43].

There are several methods to evaluate the SBH at metal/semiconductor interfaces
using first-principles calculations [14]. In this section, we adopt the standard
method using the local potential difference [44], where the SBH at the Al/InN
interface is obtained as $\text{SBH} = \left(E_{\text{F}}^{\text{b}} - V_{\text{Al}}^{\text{b}}\right) - \left(E_{\text{V,InN}}^{\text{b}} - V_{\text{InN}}^{\text{b}}\right) + \left(V_{\text{Al}}^{\text{if}} - V_{\text{InN}}^{\text{if}}\right) +$
$\left(E_{\text{V,InN}}^{\text{b}} - E_{\text{C,InN}}^{\text{b}}\right)$. Here, the first term represents the energy difference between the
Fermi energy and the average local potential in a pseudomorphic Al bulk system,
while the second term represents the energy difference between the top of the
valence band and the average local potential in bulk InN. The third term represents
the energy difference between average local potentials in Al and InN layers in the
Al/InN interface system. For the estimation of $V_{\text{Al}}^{\text{if}}$ and $V_{\text{InN}}^{\text{if}}$, we employed the inner
Al and InN layers in the present repeated slab system. The fourth term is the
band-gap energy of bulk InN, for which the observed value of 0.7 eV is adopted
[45, 46].

The atomic positions and electronic structures of metal/InN interfaces are calculated using the standard first-principles total-energy method in the generalized
gradient approximation (GGA), adopting Vanderbilt-type ultrasoft pseudopotentials
for In (3d, 4s, 4p), N (2s, 2p), Al (3s, 3p), and Au (3d, 4s, 4p) atoms [29], the
Perdew-Burke-Ernzerhof (PBE) form of exchange-correlation potential [30], a 25

ryd cutoff energy for plane-waves expansion of wave functions, and the TAPP code
[13]. The atomic positions of the hypothetical hydrogen atoms and two InN layers
on the back surface are fixed at bulk positions, while the other atomic positions are
optimized. The other calculation details are also described elsewhere [14, 25, 47].

8.3.2 Electronic Structure and Schottky Barrier

First, we consider the electronic structure of metal/InN interfaces. The calculated
results showed that there are no qualitative differences in electronic structure
between N- and In-polarity interfaces and between Al and Au interfaces, thus, we
hereafter consider the N-polarity Al/InN interface system for example.
Figure 8.15a, b show the calculated band structure of the $Al_7/(InN)_6$ and isolated
$(InN)_6$ slabs, respectively. Comparing these band structures, one can observe that
the band structure of the Al/InN interface system is roughly equal to the sum of the
band structures of the InN and Al slabs. For example, the bands located between
-17 and -14 eV in Fig. 8.15a originate from 2s orbitals of N atoms, which can also
be observed between -14 and -12 eV in Fig. 8.15b. The bands between -7 and
-2 eV in Fig. 8.15a originate from N 2p orbitals, which can also be observed
between -5 and 0 eV in Fig. 8.15b. In the case of the InN slab, both the highest
valence-band and lowest conduction-band states respectively appear at approximately 0.0 and 0.5 eV at the Γ point, as shown in Fig. 8.15b. These states are
located at approximately -2 and -1.5 eV, respectively, at Γ in Fig. 8.15a in the
case of the Al/InN interface.

The most interesting feature of the Al/InN interface system is the position of the
Fermi energy. As shown in Fig. 8.15a, the Fermi energy is located about 1.3 eV
above the lowest conduction-band state (denoted as c in the figure) of the InN

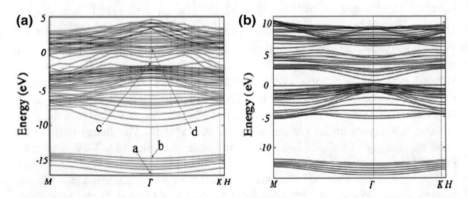

Fig. 8.15 Calculated band structures of **a** N-polarity $Al_7/(InN)_6$ interface system and **b** isolated
$(InN)_6$ slab system. Fermi energy positions are located at 0.0 eV in both **a** and **b**. Reproduced with
permission from Takei and Nakayama [41]. Copyright (2009) by the Japan Society of Applied
Physics

layers. Since the metallic Al layers mainly produce the Fermi-energy states in Al/InN system, this result indicates that a certain amount of electron-charge transfer occurs from Al to InN layers at the interface. By analyzing charge distribution and potential profile around the interface, we can confirm such electron transfer across the interface (not shown here. See [41] in details).

From these electronic structure analyses, we can describe the schematic band alignment near the Al/InN interface, which is shown in Fig. 8.16. Since the Fermi energy of Al is located above the lowest conduction band of InN, the electron transfer occurs from Al layers into the near-interface region of InN layers. This transfer promotes the band bending on a microscopic scale (10–15Å) in InN layers, which is also confirmed by observing the electronic states in InN layers (See [41]).

Next, we consider the intrinsic Schottky-barrier heights (SBHs) at metal/InN interfaces. In all N-polarity and In-polarity Al/InN and Au/InN interfaces, we observed the above-mentioned microscopic potential bending in InN layers, which is caused by the electron transfer from metal layers to InN layers as shown in Fig. 8.16. The values of intrinsic SBHs shown in Fig. 8.16 are derived by removing this effect of potential bending [41].

Table 8.1 shows the calculated intrinsic SBHs at Al/InN and Au/InN interfaces [41], where the error is estimated as about 0.1 eV by considering slabs with

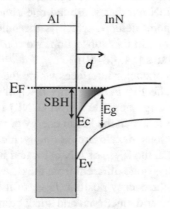

Fig. 8.16 Schematic band alignment at Al/InN interface. E_C and E_V are the lowest conduction-band and the highest valence-band edges of InN, while E_g is the band gap of InN. SBH denotes the intrinsic Schottky-barrier height for electron. Electrons transfer from Al electrode into interface region of InN having $d = 10$–15 Å width

Table 8.1 Calculated intrinsic Schottky barrier heights (SBH) at N- and In-polarity metal/InN interfaces in eV. Positive values indicate that Fermi energies of metals are located above the conduction-band bottom of InN. See also Fig. 8.16

	N polarity	In polarity
Al/InN	1.32 ± 0.1	1.23 ± 0.1
Au/InN	0.43 ± 0.1	0.62 ± 0.1

different thickness. These SBH values are also evaluated from the calculated band structure, i.e., the projected density of states. The positive values indicate that the Fermi energies of both Al and Au metals are located above the conduction-band bottom of InN, which is consistent with the band-alignment prediction of Van de Walle and Neugebauer [39]. The difference in SBH between Al and Au metals is approximately 0.7 eV in average, which originates from the intrinsic difference in the work function between Al and Au bulk metals. On the other hand, the decrease in the difference in SBH between Al and Au metals is observed at the In-polarity interface but not at the N-polarity interface, which is mainly caused by the increase in SBH at the N-polarity Au/InN interface. This occurs because Au atoms have substantially higher electronegativity than In atoms, and the Au-In bond at the interface promotes electron transfer (polarization) from In to Au, which thus decreases the electron transfer from Au to InN layers and increases SBHs.

8.3.3 Origin of Electron Carrier Generation by Structural Defects

As shown in this chapter, electron carriers are generated in conduction bands of InN at two-dimensional metal/InN interfaces, around one-dimensional dislocations, and around zero-dimensional point defects, which is the unique feature of InN never seen in other semiconductors. In this subsection, we consider this feature from the viewpoint of interface physics [14, 48–50].

For most of metal/semiconductor interfaces where the translational symmetry is broken, we can define the effective Fermi energy E_{CNL} of semiconductor in its band gap, which is called the charge neutrality level (CNL) [14]. At the interface, the charge transfer occurs so as to match the Fermi energy of metal E_F to this E_{CNL}, thus, the Fermi energies of various metals are normally located in the band gap of semiconductor. In this sense, the E_{CNL} of InN is located in the conduction bands as shown in Fig. 8.17a, which is quite different from the cases of other semiconductors.

First, we consider why the energy position E_{CNL} of InN is so different. Since In and N have large (due to inner d electrons) and small atomic radiuses, respectively, the transfer energy of In-s orbital between adjacent In atoms becomes extremely large in the tight-binding picture and thus the lowest conduction band has markedly large dispersion, i.e., large band width as seen for the band c1 in Fig. 8.8a and thus small effective mass. In addition, since these In-s orbitals receive the strong Coulomb interaction of N atoms, such conduction band almost touches the valence-band top and thus the band gap becomes so small for InN. By the way, electronic structures of most of semiconductors are produced by sp^3 orbitals and their band gaps appear as the energy gaps between bonding and antibonding states of these orbitals, which teaches us that the band gap is produced not only by s orbitals but also p orbitals. We note in Fig. 8.8a that the second lowest conduction band made of antibonding p orbitals is located about 1–1.5 eV above the lowest

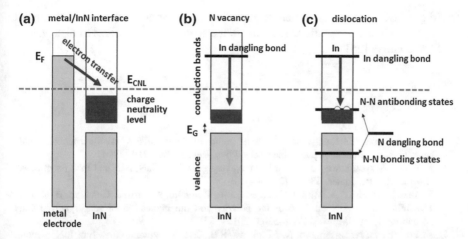

Fig. 8.17 Schematic energy diagrams explaining electron carrier generation in InN films in cases of **a** metal/InN interface, **b** N vacancy in InN, and **c** dislocation in InN. E_G indicates the band-gap position between lower valence (lower box) and upper conduction (upper box) bands, while E_F is the Fermi-energy position. Green band regions are occupied by electrons. Red arrows indicate the electron transfer in these systems, while red regions indicate the generated electron carriers in the conduction bands of InN

conduction band. The energy position of CNL is obtained as some average of energy differences of various bonding-antibonding gaps. In this sense, we can understand the suitability that the CNL of InN is located in the lowest conduction band as shown in Fig. 8.17a.

Next, we consider the case of point defect. In most of semiconductors, the point-defect-induced dangling bonds have energy levels in the band gap. In the case of InN, however, as seen in Fig. 8.17b, the dangling bond states of In atoms appear in the conduction band region and supply electron carriers into the conduction bands. Considering that the dangling bonds around point defects are often made of sp^3-like orbitals, the similar scenario to that of the above-mentioned metal/InN interfaces also applies to the case of point defects. It is interesting to note from the viewpoint of interface physics, on the other hand, that the CNL is determined by the metal-induced gap states (MIGS) in the case of plane metal/semiconductor interface, while the CNL is determined by the disorder-induced gap states (DIGS) in the case of point defects [14].

Finally, we consider the case of dislocation. In most of semiconductors such as Si and GaAs, dangling-bond-like orbitals localized around the dislocation core have energy levels in the band gap [25]. In the case of InN, however, as shown in Fig. 8.17c, since the lowest conduction band is wide, not only the In dangling-bond states but also N-N antibonding states are located in the lowest conduction band and supply electron carriers to such conduction band, thus the electron carrier generation occurs. As explained in Sect. 8.3.2, however, the latter N-N states appear in the band gap of GaN, there is no generation of electron carriers in the conduction band in the case of GaN dislocation. This situation also occurs in the cases of dislocations in Si and GaAs [25].

References

1. M. Murayama, T. Nakayama, Chemical trend of band offsets at wurtzite/zinc-blende heterocrystalline semiconductor interfaces. Phys. Rev. B **49**, 4710 (1994)
2. X. Wang, A.Yoshikawa, Molecular beam epitaxy growth of GaN, AlN and InN. Prog. Cryst. Growth Charact. Mater. **48/49,** 42 (2004)
3. T. Auzelle, B. Haas, A. Minj, C. Bougerol, J.-L. Rouvière, A. Cros, J. Colchero, B. Daudin, The influence of AlN buffer over the polarity and the nucleation of self-organized GaN nanowires. J. Appl. Phys. **117**, 245303 (2015)
4. A. Yoshikawa, N. Hashimoto, N. Kikukawa, S.B. Che, Y. Ishitani, Growth of InN quantum dots on N-polarity GaN by molecular-beam epitaxy. Appl. Phys. Lett. **86**, 153115 (2005)
5. Y. Kangawa, K. Kakimoto, T. Ito, A. Koukitu, Thermodynamic stability of $In_{1-x\ -y}Ga_xAl_yN$ on GaN and InN. Phys. Status Solidi C **3**, 1700 (2006)
6. N. Kawaguchi, K. Hida, Y. Kangawa, Y. Kumagai, A. Koukitu, Pulse laser assisted MOVPE for InGaN with high indium content. Phys. Status Solidi A **201**, 2846 (2004)
7. R. Katayama, Y. Fukuhara, M. Kakuda, S. Kuboya, K. Onabe, S. Kurokawa, N. Fujii, T. Matsuoka, Optical properties of the periodic polarity-inverted GaN waveguides. Proc. SPIE **8268**, 826814 (2012)
8. D.H. Lim, K. Xu, S. Arima, A. Yoshikawa, K. Takahashi, Polarity inversion of GaN films by trimethyl–aluminum preflow in low-pressure metalorganic vapor phase epitaxy growth. J. Appl. Phys. **91**, 6461 (2002)
9. C. Li, H. Liu, S.J. Chua, Influences of group-III source preflow on the polarity, optical, and structural properties of GaN grown on nitridated sapphire substrates by metal-organic chemical vapor deposition. J. Appl. Phys. **117**, 125305 (2015)
10. T. Nakayama, J. Mikami, Ultrathin metal layers to convert surface polarity of nitride semiconductors. Phys. Status Solidi B **242**, 1209 (2005)
11. J. Fritsch, O.F. Sankey, K.E. Schmidt, J.B. Page, *Ab initio* calculation of the stoichiometry and structure of the (0001) surfaces of GaN and AlN. Phys. Rev. B **57**, 15360 (1998)
12. K. Shiraishi, A new slab model approach for electronic structure calculation of polar semiconductor surface. J. Phys. Soc. Jpn. **59**, 3455 (1990)
13. Computer program package TAPP (Tokyo Ab-initio Program Package) and xTAPP, University of Tokyo 1983–2016
14. T. Nakayama, Y. Kangawa, K. Shiraishi, Atomic structures and electronic properties of semiconductor interfaces, in *Comprehensive Semiconductor Science and Technology*, ed. by P. Bhattacharya, R. Fomari, H. Kamimura, vol. I. (Elsevier B.V., Amsterdam, 2011), pp. 113–174
15. J.E. Northrup, J. Neugebauer, R.M. Feenstra, A.R. Smith, Structure of GaN(0001): The laterally contracted Ga bilayer model. Phys. Rev. B **61**, 9932 (2000)
16. Y. Kangawa, T. Akiyama, T. Ito, K. Shiraishi, T. Nakayama, Surface stability and growth kinetics of compound semiconductors: an Ab initio-based approach. Materials **6**, 3309 (2013)
17. T. Harumoto, T. Sannomiya, Y. Matsukawa, S. Muraishi, J. Shi, Y. Nakamura, H. Sawada, T. Tanaka, Y. Tanishiro, K. Takayanagi, Controlled polarity of sputter-deposited aluminum nitride on metals observed by aberration corrected scanning transmission electron microscopy. J. Appl. Phys. **113**, 084306 (2013)
18. T. Nakayama, Y. Takei, Surface strain and hexagonal/cubic polymorphism in InGaN epitaxy: first-principles study. Phys. Status Solidi C **4**, 259 (2007)

19. S.K. Hong, T. Hanada, H.J. Ko, Y. Chen, T. Yao, D. Imai, K. Araki, M. Shinohara, K. Saitoh, M. Terauchi, Control of crystal polarity in a wurtzite crystal: ZnO films grown by plasma-assisted molecular-beam epitaxy on GaN. Phys. Rev. B **65**, 115331 (2002)
20. M. Adachi, M. Takasugi, M. Sugiyama, J. Iida, A. Tanaka, H. Fukuyama, Polarity inversion and growth mechanism of AlN layer grown on nitride sapphire substrate using Ga–Al liquid-phase epitaxy. Phys. Status Solidi B **252**, 743 (2015)
21. Y. Nanishi, Y. Saito, T. Yamauchi, RF-molecular beam epitaxy growth and properties of InN and related alloys. Jpn. J. Appl. Phys. **42**, 2549 (2003)
22. J. Wu, W. Walukiewicz, Band gaps of InN and group III nitride alloys. Superlattices Microstruct. **34**, 63 (2003)
23. I. Mahboob, T.D. Veal, C.F. McConville, H. Lu, W.J. Schaff, Intrinsic electron accumulation at clean InN surfaces. Phys. Rev. Lett. **92**, 036804 (2004)
24. X. Wang, S.B. Che, Y. Ishitani, A. Yoshikawa, Threading dislocations in In-polar InN films and their effects on surface morphology and electrical properties. Appl. Phys. Lett. **90**, 151901 (2007)
25. R. Kobayashi, T. Nakayama, Atomic and electronic structures of stair-rod dislocations in Si and GaAs. Jpn. J. Appl. Phys. **47**, 4417 (2008)
26. Y. Takei, T. Nakayama, Electron carrier generation at edge dislocations in InN films; first-principles study. J. Cryst. Growth **311**, 2767 (2009)
27. A. Béré, A. Serra, Atomic structure of dislocation cores in GaN. Phys. Rev. B **65**, 205323 (2002)
28. H.P. Lei, P. Ruterana, G. Nouet, X.Y. Jiang, J. Chen, Core structures of the aa-edge dislocation in InN. Appl. Phys. Lett. **90**, 111901 (2007)
29. D. Vanderbilt, Soft self-consistent pseudopotentials in a generalized eigenvalue formalism. Phys. Rev. B **41**, 7892 (1990)
30. J.P. Perdew, K. Burke, M. Ernzerhof, Generalized gradient approximation made simple. Phys. Rev. Lett. **77**, 3865 (1996)
31. Y. Ishitani, W. Terashima, S.B. Che, A. Yoshikawa, Conduction and valence band edge properties of hexagonal InN characterized by optical measurements. Phys. Status Solidi C **3**, 1850 (2006)
32. M. Murayama, T. Nakayama, *Ab initio* calculations of two-photon absorption spectra in semiconductors. Phys. Rev. B **52**, 4986 (1995)
33. A. Yoshikawa, X. Wang, Y. Ishitani, A. Uedono, Recent advances and challenges for successful p-type control of InN films with Mg acceptor doping by molecular beam epitaxy. Phys. Status Solidi A **207**, 1011 (2010)
34. Y. Takei, "Electronic structures of surfaces, interfaces, and defects of InN nitride semiconductor" (in Japanese), PhD Thesis, Chiba University (2009)
35. A. Faghaninia, J.W. Ager III, C.S. Lo, *Ab initio* electronic transport model with explicit solution to the linearized Boltzmann transport equation. Phys. Rev. B **91**, 235123 (2015)
36. N. Miller, E.E. Haller, G. Koblmüller, C. Gallinat, J.S. Speck, W.J. Schaff, M.E. Hawkridge, K.M. Yu, J.W. Ager III, Effect of charged dislocation scattering on electrical and electrothermal transport in n-type InN. Phys. Rev. B **84**, 075315 (2011)
37. E. Baghani, S.K. O'Leary, Electron mobility limited by scattering from screened positively charged dislocation lines within indium nitride. Appl. Phys. Lett. **99**, 262106 (2011)
38. N. Miller, J.W. Ager III, H.M. Smith III, M.A. Mayer, K.M. Yu, E.E. Haller, W. Walukiewicz, W.J. Schaff, C. Gallinat, G. Koblmüller, J.S. Speck, Hole transport and photoluminescence in Mg-doped InN. J. Appl. Phys. **107**, 113712 (2010)
39. C.G. Van de Walle, J. Neugebauer, Universal alignment of hydrogen levels in semiconductors, insulators and solutions. Nature **423**, 626 (2003)
40. J. Tersoff, Schottky barrier heights and the continuum of gap states. Phys. Rev. Lett. **52**, 465 (1984)
41. Y. Takei, T. Nakayama, First-principles study of Schottky-Barrier behavior at metal/InN interfaces. Jpn. J. Appl. Phys. **48**, 081001 (2009)

42. T. Nakayama, S. Itaya, D. Murayama, Nano-scale view of atom intermixing at metal/ semiconductor interfaces. J. Phys: Conf. Ser. **38**, 216 (2006)
43. T. Nakayama, K. Shiraishi, S. Miyazaki, Y. Akasaka, K. Torii, P. Ahmet, K. Ohmori, N. Umezawa, H. Watanabe, T. Chikyow, Y. Nara, A. Ohta, H. Iwai, K. Yamada, T. Nakaoka, Physics of Metal/High-k Interfaces. ECS Trans. **3**, 129 (2006). and references therein
44. T. Nakayama, Valence band offset and electronic structures of zinc-compound strained superlattices. J. Phys. Soc. Jpn. **61**, 2434 (1992)
45. D. Muto, H. Naoi, T. Araki, S. Kitagawa, M. Kurouchi, H. Na, Y. Nanishi, High-quality InN grown on KOH wet etched N-polar InN template by RF-MBE. Phys. Status Solidi A **203**, 1691 (2006)
46. Y. Ishitani, W. Terashima, S.B. Che, A. Yoshikawa, Conduction and valence band edge properties of hexagonal InN characterized by optical measurements. Phys. Status Solidi C **3**, 1850 (2006)
47. S. Sakurai, T. Nakayama, Electronic structures and etching processes of Chlorinated Si(111) Surfaces. Jpn. J. Appl. Phys. **41**, 2171 (2002)
48. S. Picozzi, A. Continenza, G. Satta, S. Massidda, A.J. Freeman, Metal-induced gap states and Schottky barrier heights at nonreactive GaN/noble-metal interfaces. Phys. Rev. B **61**, 16736 (2000)
49. J.W. Ager III, N.R. Miller, Taming transport in InN. Phys. Status Solidi A **209**, 83 (2012)
50. N. Spyropoulos-Antonakakis, E. Sarantopoulou, Z. Kollia, G. Dražic, S. Kobe, Schottky and charge memory effects in InN nanodomains. Appl. Phys. Lett. **99**, 153110 (2011)

Chapter 9
Defects in Indium-Related Nitride Compounds and Structural Design of AlN/GaN Superlattices

Kenji Shiraishi

In this chapter, we focus on two topics related to the electronic and optical properties of III-nitride compounds. By applying of ab initio approach, we can analyze the electronic structures of III-nitride compounds as well as other semiconductors. This is exemplified by theoretical analysis of electronic structures of In-related nitride compounds, which exhibit characteristic behavior originating from the large difference in the covalent radius between In and N atoms. By considering atomic and electronics structures of nitrogen vacancy (V_N) in InGaN in detail, the second nearest neighbor In–In interaction are crucial for unusually narrow bandgap of InN. Furthermore, this approach is applied to demonstrate AlN/GaN superlattice in the wurtzite phase with one or two GaN monolayers, which is efficient for near-band-edge c-plane emission of deep-ultraviolet (UV) LEDs. In particular, the emission wavelength is estimated to be 224 nm for the AlN/GaN superlattice with one GaN-monolayer, which is remarkably shorter than that for Al-rich AlGaN alloys. The optical matrix element of such superlattice is found to be 57% relative to the GaN bulk value. In Sect. 9.1, the atomic and electronic structures of V_N in InGaN to clarify the physical origin of the unusually narrow bandgap of InN. Section 9.2 is devoted to discuss structural design of AlN/GaN superlattices for deep-ultraviolet light-emitting diodes with high emission efficiency.

9.1 Defects in Indium-Related Nitride Compounds

In-related nitride compounds such as InGaN have attracted great attention from both technological and scientific viewpoints since InGaN is applied as a key material for blue light-emitting diodes [1]. Although it has been practically used in

K. Shiraishi (✉)
Institute of Materials and Systems for Sustainability, Nagoya University, Nagoya, Japan
e-mail: shiraishi@cse.nagoya-u.ac.jp

© Springer International Publishing AG, part of Springer Nature 2018   171
T. Matsuoka and Y. Kangawa (eds.), *Epitaxial Growth of III-Nitride Compounds*,
Springer Series in Materials Science 269,
https://doi.org/10.1007/978-3-319-76641-6_9

the semiconductor industry, many fundamental phenomena observed in In-related nitride compounds are still mysterious. The physical origin of the unusually narrow bandgap of InN [2–4] has not been clearly explained: the very small bandgap has been obtained in previous ab initio calculations [5]. In order to clarify the origin of the peculiar properties, it should be noted that the covalent radius of an In atom (1.44 Å) is much larger than that of an N atom (0.7 Å). This characteristic feature results in the relatively shorter distance between the second nearest neighbor In atoms as schematically shown in Fig. 9.1. As shown in this figure, the In–In distance in InN (3.5 Å) is much smaller than that in the typical covalent semiconductor of InAs (4.28 Å) due to the small covalent radius of N. It should also be noted that the In–In distance in a bulk In metal is 3.25 Å. These structural characteristics imply that the second nearest neighbor In–In interaction is of importance in In-related nitride compounds. It is thus expected that V_N in InGaN should reflect the importance of the In–In interaction.

The analysis of electronic structures of V_N for various atomic configuration in InGaN demonstrates that the second nearest neighbor In–In interaction is crucial as well as the nearest neighbor In–N interaction in In-related nitride compounds. Figure 9.2 shows the schematics of atomic configurations of V_N in InGaN considered in ab initio calculations. The calculated formation energy of V_N, which is defined as the minimum energy cost to generate V_N by removing N atom in InGaN, [6, 7] reveals that the formation energy of positively charged V_N is lower than that with neutral charge state. The calculated formation energy of V_N in n-type GaN is 2.84 eV under Ga-rich condition, where the chemical potential of N is assumed using (4.2), agrees with that in previous calculations [8]. Table 9.1 shows the relative formation energies of positively charged V_N. The formation energy in pure GaN is set to be zero for comparison. It is found that the formation energy decreases with the number of neighboring In atoms. The single substitution of In atom causes the decrease in the formation energy by ~ 0.5 eV. Consequently, substitution of two In atoms for two Ga atoms results in the 1.0 eV decrease in the formation

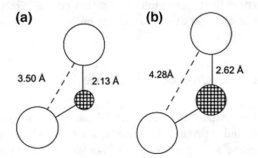

Fig. 9.1 Schematic illustration of the importance of the second nearest neighbor by considering the strong second nearest neighbor In–In interactions in In-related nitride semiconductors (**a**) compared with a conventional semiconductor of InAs (**b**). The open, small hatched and large hatched circles indicate In, N, and As atoms, respectively. Reproduced with permission from Obata et al. [12]. Copyright (2009) by Elsevier

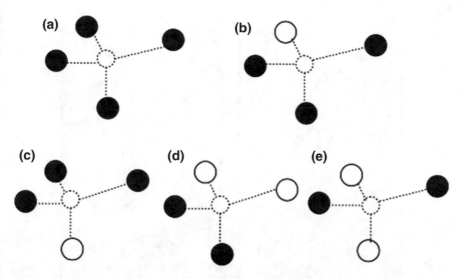

Fig. 9.2 Schematic illustration of the five investigated N mono-vacancy (V_N) structures. **a** V_N in perfect GaN, **b** V_N with one nearest neighbor In atom (Type I), **c** V_N with one nearest neighbor In atom (Type II), and **d** V_N with two nearest neighbor In atoms (Type I). The open and filled circles indicate In and Ga atoms, respectively and the open circle with the broken line corresponds to N mono-vacancies. Reproduced with permission from Obata et al. [12]. Copyright (2009) by Elsevier

Table 9.1 Relative formation energies of V_N in InGaN compared with the V_N formation energy in pure GaN. Atomic configuration of V_N in InGaN are shown in Fig. 9.2

Type of V_N	Formation energy (eV)
Pure GaN	0.0
One In (Type I)	−0.57
One In (Type II)	−0.50
Two In (Type I)	−1.05
Two In (Type II)	−0.96

energy of V_N. In covalent semiconductors, the formation energy of the vacancy is generally discussed in terms of a dangling bond counting model [9, 10]. The formation energy corresponds to the energy cost to disrupt the four bonds around the vacancy. Therefore, it is considered that the decrease in the formation energy corresponds to the energy difference between Ga N and In–N. However, the energy gains by the formation of Ga–N and In–N bonds are estimated as 2.26 and 1.99 eV, respectively [11]. The estimated value of the decrease of the formation energy is (0.27 eV) is only half of the calculated value (~0.5 eV). This indicates that additional energy gain of 0.23 eV are important in InGaN

To clarify the origin of the above discrepancy, the atomic structures of V_N with two nearest neighbor In atoms have been investigated. Figure 9.3a shows the schematics of atomic positions of In and N atoms around V_N along with the calculated contour plots of total charge density of InGaN with V_N [12]. As shown in

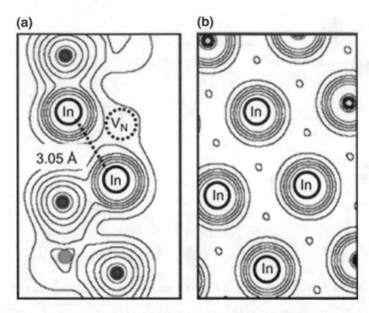

Fig. 9.3 Contour plots of total charge density. **a** InGaN with V_N, and **b** bulk In metal. The open circle with solid line corresponds to In atoms and the open circle with broken line indicates N mono-vacancies. Each contour represents twice (or half) of the density of the adjacent contour lines. The lowest values represented by the contour is 0.11 Å. Reproduced with permission from Obata et al. [12]. Copyright (2009) by Elsevier

Fig. 9.3a, the calculated In–In distance, which is originally the second nearest neighbor atoms, is only 3.05 Å. This value is even smaller than the In–In distance in bulk In (3.25 Å), indicating a strong interaction between the second nearest neighbor In–In atoms. In fact, the total charge density shown in Fig. 9.3a reflects the strong interaction between In atoms. From the calculated contour plots bulk In shown Fig. 9.3b, it is also found that the In–In interaction in InGaN with V_N is much stronger than that in bulk In, i.e., the In–In bond generated around V_N is much stronger than the In–In bond in bulk In, indicating that the concept of dangling bonds which is relevant in the conventional semiconductors is insufficient in InGaN. The formation of strong In–In bond is qualitatively consistent with theoretical reports that multiple N vacancies in InN cause metallic bonding between In–In atoms in InN [13]. Furthermore, the formation metallic In–In bonds is estimated to be 0.4 eV. This value is sufficient to explain above mentioned the additional energy gain of 0.23 eV. Moreover, the energy gain reduction of 0.17 eV seems to be natural by considering the fact that In–In bond formation in InGaN induces energetically unfavorable lattice distortion around V_N. Therefore, it is necessary to adopt a new material concept that the second nearest neighbor interaction between In atoms is very important as well as the nearest neighbor interaction for considering In-related nitride compounds.

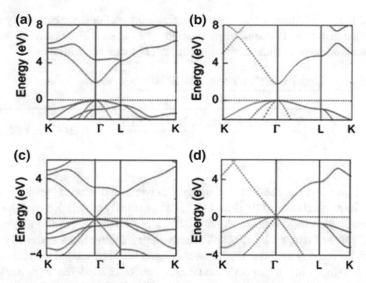

Fig. 9.4 Band structures of **a** wurtzite GaN, **b** zinc-blende GaN, **c** wurtzite InN, and **d** zinc-blende InN. The energy of valence band top is set to zero. Reproduced with permission from Obata et al. [12]. Copyright (2009) by Elsevier

By considering the second nearest neighbor In–In interactions, the physical origin of the unusually narrow bandgap of InN can be explained. In the wurtzite structure, each In atom is surrounded by 4 nearest neighbor N atoms and 12 nearest neighbor In atoms. Figure 9.4 shows the band structures of GaN and InN with both wurtzite and zinc-blende structures. As shown in these figures, InN bandgaps are much smaller than GaN for both the wurtzite and zinc-blende structures. It is noticeable that InN bandgaps become smaller due to the large dispersions of the conduction bands (especially around Γ point) for both wurtzite and zinc blende structures. This large dispersion corresponds to the very small electron effective mass of InN. Since both the wurtzite and zinc blende band structures have similar characteristics, and since the atomic configurations of the wurtzite and zinc blende structures are the same within the second nearest neighbor atoms, it is sufficient to just consider the zinc blende band structure in order to clarify the physical origin of the narrow bandgap in InN. In zinc-blende structures, the energy level at the Γ point can be described analytically based on the tight-binding calculations [14]. If we consider only the nearest neighbor interactions, the energy of the conduction band bottom at the Γ point, $E(\Gamma^c)$, is expressed as

$$E(\Gamma^c) = \frac{E(\text{In}, s) + E(\text{N}, s)}{2} + \sqrt{\left(\frac{E(\text{In}, s) - E(\text{N}, s)}{2}\right)^2 + 16t^2(\text{In}, s : \text{N}, s)}, \quad (9.1)$$

where, $E(\text{In}, s)$ and $E(\text{N}, s)$ are the on-site energies of the In-5s and N-2s orbitals, respectively, and $t(\text{In}, s : \text{N}, s)$ is the transfer integral between In-5s and

N-2s orbitals. In addition to the nearest neighbor interactions, we can easily include the interaction between second nearest neighbor In atoms. The obtained energy of the conduction band bottom at the Γ point, $E'(\Gamma^c)$ can be written as

$$
\begin{aligned}
E'(\Gamma^c) = {} & \frac{E(\text{In}, s) + 12t(\text{In}, s : \text{In}, s) + E(\text{N}, s)}{2} \\
& + \sqrt{\left(\frac{E(\text{In}, s) + 12t(\text{In}, s : \text{In}, s) - E(\text{N}, s)}{2}\right)^2 + 16t^2(\text{In}, s : \text{N}, s)},
\end{aligned}
\tag{9.2}
$$

where $t(\text{In}, s : \text{In}, s)$ is the transfer integral between the second nearest neighbor In-5s orbitals. As shown in (9.1) and (9.2), it is naturally concluded that the energy of the conduction band bottom should drastically decrease when the magnitude of negative value of $t(\text{In}, s : \text{In}, s)$ is sufficiently large, leading to an extremely narrow bandgap. To roughly estimate the transfer integral between second nearest neighbor In atoms, $t(\text{In}, s : \text{In}, s)$, for InN and InAs, we calculated the hypothetical In_2 molecules whose distances correspond to InN (3.50 Å) and InAs (4.28 Å), respectively. The calculated values of bonding–antibonding splittings of In-5s orbitals are 1.2 and 0.4 eV for 3.50 and 4.28 Å, respectively. Therefore, it is expected that $t(\text{In}, s : \text{In}, s)$ of InN is three times larger than that of InAs. This implies that second nearest In–In interaction is much more significant in InN than that in InAs. Moreover, it is also concluded that the decrease in the conduction band energy which originates from the second nearest neighbor In–In interaction is three times larger in InN than that in InAs. Since the strong In–In interaction governs the band width, it is concluded that the band dispersion of the conduction band should be large. The unusually narrow bandgap of InN can naturally be explained by considering the strong second nearest neighbor In–In interactions.

9.2 Structural Design of AlN/GaN Superlattices for Deep-UV LEDs with High Emission Efficiency

The UV LEDs have recently attracted a great deal of attention as promising candidates for the next-generation high-density optical storage and solid-state lighting. Recently, fabrication of deep-UV AlN LEDs has been reported with an emission efficiency of 1×10^{-4} % [15–21]. However, this emission efficiency is still much lower than that of the near-UV GaN LED. This is because AlN has an anisotropic emission pattern where light is emitted barely from the (0001) planes (c-plane), but preferentially from the $(11\bar{2}0)$ planes (a-plane). An increase in its emission efficiency is thus necessary for practical application of deep-UV AlN LEDs.

The directional light emission properties of AlN LEDs reflect a different valence band structure for AlN compared with that of GaN, as described in Sect. 3.2

[22–24] In general, the valence bands at the Γ point for wurtzite crystals such as AlN and GaN are separated into heavy-hole (HH), light-hole (LH), and crystal-field-split-off hole (CH) bands due to spin-orbit interactions and the crystal field. The HH and LH states have a character of nitrogen 2p orbitals perpendicular to the c-axis, while the CH state is characterized by nitrogen 2p orbitals parallel to the c-axis. The optical transitions between the HH or LH bands and the conduction band are mainly allowed for light polarized perpendicular to the c-axis ($\mathbf{E} \perp$c), where $\mathbf{E}$ is the electric field vector of the emitted light. On the other hand, the transition between the CH band and the conduction band is predominantly allowed for light polarized parallel to the c-axis ($\mathbf{E} \parallel$ c).

In GaN, the HH and LH bands are higher in energy than the CH with crystal-field splitting Δ_{cf} of ~ 25 meV [25, 26]. By contrast, in AlN, the CH band is energetically higher than the HH and LH bands, which leads to negative crystal-field splitting Δ_{cf} of -165 meV [21]. The difference in electronic structure between AlN and GaN results in the anisotropic emission pattern of the AlN. Therefore, to increase the efficiency of the c-plane surface emission in deep-UV AlN LEDs, the important issue is how to convert the negative Δ_{cf} to positive. For this purpose, the conventional method is using the ternary alloy of AlGaN. It has been reported, however, that the Ga composition at which Δ_{cf} becomes zero is 22–27%, [27] which limits its c-plane emission wavelength to ~ 240–260 nm. To obtain near-band-edge c-plane emission of deep-UV LEDs, we consider a [0001]-oriented AlN/GaN superlattice in the wurtzite phase. The period of the superlattice along the [0001] direction (c-axis) was fixed to be 20 monolayers, i.e., $(AlN)_{20-n}/$ $(GaN)_n$, where n is the number of monolayers ($n = 1, \ldots, 10$), is taken into account as a representative of [0001]-oriented AlN/GaN superlattice. Figure 9.5 shows the structural model of the AlN/GaN superlattices.

Figure 9.6 shows the electronic energy bands of the $(AlN)_{20-n}/(GaN)_n$ superlattices around the Γ point. In the AlN bulk, corresponding to $n = 0$, the CH band is energetically higher than the HH and LH bands, which leads to a negative of -0.23 eV [28]. It should be noted that the HH and LH bands are degenerate at the Γ point because the spin-orbit coupling is neglected. In the AlN/GaN superlattices, the HH and LH bands are, however, higher in energy than the CH band, resulting in the valence band top. The calculated Δ_{cf} as a function of the number of GaN monolayers shown in Fig. 9.7 [28] exhibits the reverse of the Δ_{cf} sign for AlN/GaN superlattices with more than one GaN monolayer. The calculated Δ_{cf} is about $+0.18$ eV in the $(AlN)_{19}/(GaN)_1$ superlattice. Furthermore, the value of Δ_{cf} increases with GaN thickness, but it approaches $+0.40$ eV in the AlN/GaN

Fig. 9.5 Structural model of AlN/GaN superlattices. The figure shows $(AlN)_{18}/(GaN)_2$ as an example. Yellow, black, and gray circles represent Al, N, and Ga atoms, respectively. Reproduced with permission from Kamiya et al. [28]. Copyright (2012) by the Japan Society of Applied Physics

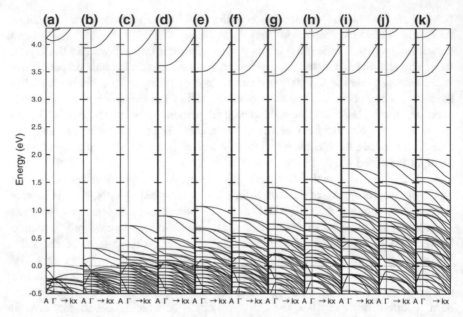

Fig. 9.6 Energy band structure around the Γ point of the **a** bulk AlN,c Energies are measured from the valence band top of bulk AlN at the Γ point. Reproduced with permission from Kamiya et al. [28]. Copyright (2012) by the Japan Society of Applied Physics

Fig. 9.7 Calculated crystal-field splitting Δ_{cf} of a AlN/GaN superlattice as a function of the number of GaN monolayers. Reproduced with permission from Kamiya et al. [28]. Copyright (2012) by the Japan Society of Applied Physics

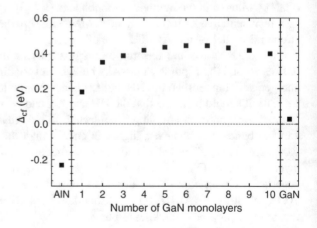

superlattices with more than three GaN monolayers. These results indicate clearly that a negative value of Δ_{cf} in the AlN bulk is converted to a positive value of Δ_{cf} by using the AlN/GaN superlattice with more than one GaN monolayer.

Figure 9.8 shows the wave functions of the lowest conduction band, HH/LH bands, and CH band for the $(AlN)_{20-n}/(GaN)_n$ superlattices [28]. It is found that the wave functions in the HH and LH bands are characterized by N 2p orbitals

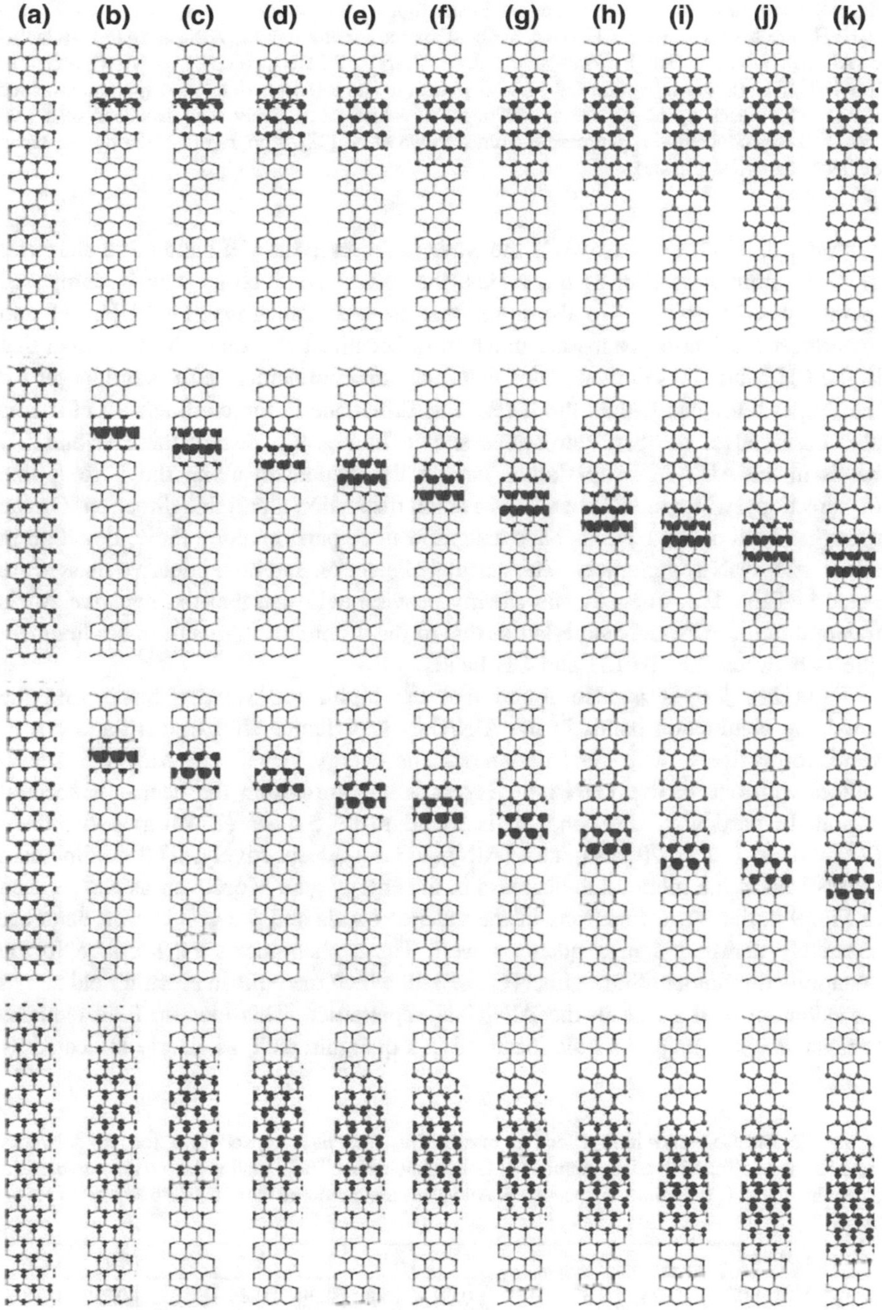
(a)
(b)
(c)
(d)
(e)
(f)
(g)
(h)
(i)
(j)
(k)

◄**Fig. 9.8** Isosurfaces of the wavefunctions in the (upper) lowest conduction band, (two middle) HH/LH bands, and (lower) CH band at the Γ point for the **a** bulk AlN, **b** $(AlN)_{19}/(GaN)_1$, **c** $(AlN)_{18}/(GaN)_2$, **d** $(AlN)_{17}/(GaN)_3$, **e** $(AlN)_{16}/(GaN)_4$, **f** $(AlN)_{15}/(GaN)_5$, **g** $(AlN)_{14}/(GaN)_6$, **h** $(AlN)_{13}/(GaN)_7$, **i** $(AlN)_{12}/(GaN)_8$, **j** $(AlN)_{11}/(GaN)_9$, and **k** $(AlN)_{10}/(GaN)_{10}$ superlattices. Blue and red isosurfaces represent positive and negative values, respectively. The isovalues are ± 1.08 $(e/Å^3)^{1/2}$. Reproduced with permission from Kamiya et al. [28]. Copyright (2012) by the Japan Society of Applied Physics

perpendicular to the c-axis, while the wave function in the CH band has a character of N-2p orbitals parallel to the c-axis; the lowest conduction state is composed mainly of N-2s orbitals. We also found that the wave functions in the HH, LH, and lowermost conduction bands are much more localized in a quantum well than that in the CH band. This demonstrates that quantum confinement effects are larger for the former three bands than the latter. This difference in the confinement effects is also observed in the band structures shown in Fig. 9.6, where the HH and LH bands in the AlN/GaN superlattice have a flat dispersion along the Γ to A line (k_z direction), while the CH band has a small dispersion along this direction. On the other hand, all of these bands have a significant dispersion along the k_x direction in all the AlN/GaN superlattices. The calculated electron and hole effective masses are listed in Table 9.2. These results clearly provide evidence that the negative Δ_{cf} is reversed in the AlN/GaN superlattice due to the difference in quantum confinement effects between the HH/LH and CH bands.

From Fig. 9.6, it is also found that the high-lying valence bands and the low-lying conduction bands in the AlN/GaN superlattice shift upward and downward, respectively, with GaN thickness. The energy gap of the AlN/GaN superlattices with one to five GaN monolayers is ranging within the deep-UV spectral region. In particular, the bandgap is found to be 5.5 eV (224 nm) and 5.0 eV (247 nm) in $(AlN)_{19}/(GaN)_1$ and $(AlN)_{18}/(GaN)_2$, respectively. The confinement effects lead to the monotonic decrease in the energy gap. Moreover, as can be seen in Fig. 9.8, the wave functions in the valence bands and the conduction band are separately distributed in a quantum well. These phenomena are ascribed to the quantum-confinement Stark effect (QCSE), in which the built-in electric field across the quantum well exists in the AlN/GaN superlattice. This electric field tends to separate an electron and a hole localized in a quantum well, as clearly indicated by

Table 9.2 Electron and hole effective masses (m_e and m_h, respectively) for an AlN/GaN superlattice as a function of the number of GaN monolayers. The $\perp$ and $\parallel$ denote the components along the k_x and k_x directions, respectively. All values are measured in units of a free-electron mass m_0

	AlN	1	2	3	4	5	6	7	8	9	10	GaN
$m_e^\perp$	0.31	0.33	0.33	0.32	0.30	0.30	0.29	0.29	0.29	0.29	0.29	0.18
$m_e^\parallel$	0.30											0.17
$m_h^\perp$	4.02	1.70	1.71	1.73	1.74	1.74	1.76	1.76	1.78	1.78	1.79	2.29
$m_h^\parallel$	0.25											2.23

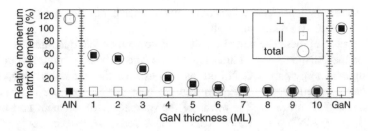

Fig. 9.9 Interband optical momentum matrix elements of an AlN/GaN superlattice as a function of GaN monolayer thickness, calculated at the C point between the highest valence band and the lowest conduction band. All the values are normalized to the square value of the momentum matrix element for GaN bulk. The $\perp$ and $\parallel$ denote the components that are perpendicular and parallel to the c-axis, respectively. Reproduced with permission from Kamiya et al. [29]. Copyright (2011) by American Institute of Physics

the distribution of the square of the wave functions in the valence band top and conduction band bottom in Fig. 9.8. It is thus suggested that the AlN/GaN super-lattice with one or two GaN monolayers is efficient for near-band-edge emission of a deep-UV LED. It should be noted that using different AlN thicknesses for the AlN/GaN superlattice, i.e., $(AlN)_{29}/(GaN)_1$ and $(AlN)_{39}/(GaN)_1$, gives the same results as with $(AlN)_{19}/(GaN)_1$: The results are independent of AlN thickness.

Figure 9.9 shows the interband optical momentum matrix element of the AlN/GaN superlattice as a function of Ga monolayer thickness, also shown in AlN bulk and GaN bulk cases; these values are shown as a relative value to that of GaN, $r_{superlatice/GaN}$, given by

$$r_{superlattice/GaN} = \frac{|\psi_c|\mathbf{P}|\psi_v|_{AlN/GaN}}{|\psi_c|\mathbf{P}|\psi_v|_{GaN}}, \tag{9.3}$$

where ψ_v and ψ_c are the wave functions of the valence band top and conduction band bottom at the Γ point, respectively, $\mathbf{P}$ is the momentum operator, and sub-scripts AlN/GaN and GaN in momentum matrix elements denote AlN/GaN superlattice and bulk GaN, respectively [29]. In the case of the AlN bulk, the component of the matrix element perpendicular to the c-axis is almost equal to zero, and the matrix element component parallel to the c-axis accounts for a majority of the total. The trend is completely opposite in the case of the GaN bulk. However, the component of the matrix element perpendicular to the c-axis significantly increases for the AlN/GaN superlattices with one and two GaN monolayers, as shown in Fig. 9.9. The calculated values are 57 and 52% relative to GaN bulk for the $(AlN)_{19}/(GaN)_1$ and $(AlN)_{18}/(GaN)_2$ cases, respectively. On the other hand, the component parallel to the c-axis is almost down to zero in all of the AlN/GaN superlattices. The momentum matrix elements decrease monotonically as the number of GaN monolayers increases. This monotonic decrease originates from the

QCSE. These results offer the improvement of near-band-edge emission from the c-plane by using AlN/GaN superlattices.

On the basis of these calculated results, it is concluded that quantum confinement effects of the AlN/GaN superlattices with one and two GaN monolayers lead to spreading the energy gap of GaN bulk as well as reversing the negative Δ_{cf} in AlN bulk: such a superlattice structure makes the best use of both the high emission efficiency of GaN up to 57% in terms of the optical matrix element and the wide energy gap of AlN up to 5.5 eV (224 nm). Since 22–27% Ga composition in ternary alloy of AlGaN is needed to reverse the Δ_{cf} sign, [27] which limits the emission wavelength to ~ 240–260 nm, this is the advantageous of using superlattices compared with using the ternary alloy of AlGaN.

References

1. S. Nakamura, M. Senoh, S. Nagahama, N. Iwasa, T. Yamada, T. Matsushita, H. Kiyoku, Y. Sugimoto, InGaN-based multi-quantum-well-structure laser diodes. Jpn. J. Appl. Phys. **37**, L74 (1996)
2. T. Matsuoka, H. Okamoto, M. Nakao, H. Harima, E. Kurimoto, Optical bandgap energy of wurtzite InN. Appl. Phys. Lett. **81**, 1246 (2002)
3. V.Y. Davydov, A.A. Klochikhin, V.V. Emtsev, S.V. Ivanov, V.V. Vekshin, F. Bechstedt, J. Fürthmuller, H. Harima, A.V. Mudryi, A. Hashimoto, A. Yamamoto, J. Aderhold, J. Graul, E.E. Haller, Band gap of InN and In-rich $In_xGa_{1-x}N$ alloys ($0.36 < x < 1$). Phys. Status Solidi B **230**, R4 (2002)
4. Y. Nanishi, Y. Saito, T. Yamaguchi, RF-molecular beam epitaxy growth and properties of InN and related alloys. Jpn. J. Appl. Phys. **42**, 2549 (2003)
5. M. Usuda, N. Hamada, K. Shiraishi, A. Oshiyama, Band structures of wurtzite InN and $Ga_{1-x}In_xN$ by all-electron GW calculation. Jpn. J. Appl. Phys. **43**, L407 (2004)
6. M. Otani, K. Shiraishi, A. Oshiyama, First-principles calculations of boron-related defects in SiO_2. Phys. Rev. B. **68**, 184112 (2003)
7. D.J. Chadi, K.J. Chang, Magic numbers for vacancy aggregation in crystalline Si. Phys. Rev. B **38**, 1523 (1988)
8. C.G. Van de Walle, J. Neugebauer, First-principles calculations for defects and impurities: Applications to III-nitrides. J. Appl. Phys. **95**, 3851 (2004)
9. T. Akiyama, A. Oshiyama, O. Sugino, Magic numbers of multivacancy in crystalline Si: tight-binding studies for the stability of the multivacancy. J. Phys. Soc. Jpn. **67**, 4110 (1998)
10. T. Akiyama, A. Oshiyama, First-principles study of hydrogen incorporation in multivacancy in silicon. J. Phys. Soc. Jpn. **70**, 1627 (2001)
11. M. Fuchs, J.L.F. DaSilva, C. Stampfl, J. Neugebauer, M. Scheffler, Cohesive properties of group-III nitrides: A comparative study of all-electron and pseudopotential calculations using the generalized gradient approximation. Phys. Rev. B **65**, 245212 (2002)
12. T. Obata, J.-I. Iwata, K. Shiraishi, A. Oshiyama, First principles studies on In-related nitride compounds. J. Cryst. Growth **311**, 2772 (2009)
13. X.M. Duan, C. Stampfl, Nitrogen vacancies in InN: vacancy clustering and metallic bonding from first principles. Phys. Rev. B **77**, 115207 (2008)
14. K.E. Newman, J.D. Dow, Theory of deep impurities in silicon-germanium alloys. Phys. Rev. B **30**, 1929 (1984)
15. T. Nishida, N. Kobayashi, 346 nm emission from AlGaN multi-quantum-well light emitting diode. Phys. Status Solidi A **176**, 45 (1999)

16. V. Adivarahan, W.H. Sun, A. Chitnis, M. Shatalov, S. Wu, H.P. Maruska, M.A. Khan, 250 nm AlGaN light-emitting diodes. Appl. Phys. Lett. **85**, 2175 (2004)
17. M.A. Khan, M. Shatalov, H.P. Maruska, H.M. Wang, E. Kuokstis, III–nitride UV devices. Jpn. J. Appl. Phys. **44**, 7191 (2005)
18. Y. Taniyasu, M. Kasu, T. Makimoto, An aluminium nitride light-emitting diode with a wavelength of 210 nanometres. Nature (London) **441**, 325 (2006)
19. A.A. Yamaguchi, Anisotropic optical matrix elements in strained GaN quantum wells on semipolar and nonpolar substrates. Jpn. J. Appl. Phys. **46**, L789 (2007)
20. A.A. Yamaguchi, Valence band engineering for remarkable enhancement of surface emission in AlGaN deep-ultraviolet light emitting diodes. Phys. Status Solidi C **5**, 2364 (2008)
21. Y. Taniyasu, M. Kasu, Origin of exciton emissions from an AlN p-n junction light-emitting diode. Appl. Phys. Lett. **98**, 131910 (2011)
22. M. Suzuki, T. Uenoyama, A. Yanase, First-principles calculations of effective-mass parameters of AlN and GaN. Phys. Rev. B **52**, 8132 (1995)
23. S.-H. Wei, A. Zunger, Valence band splittings and band offsets of AlN, GaN, and InN. Appl. Phys. Lett. **69**, 2719 (1996)
24. K. Kim, W.R.L. Lambrecht, B. Segall, M. van Schilfgaarde, Effective masses and valence-band splittings in GaN and AlN. Phys. Rev. B **56**, 7363 (1997)
25. D.C. Reynolds, D.C. Look, W. Kim, Ö. Aktas, A. Botchkarev, A. Salvador, H. Morkoç, D.N. Talwar, Ground and excited state exciton spectra from GaN grown by molecular-beam epitaxy. J. Appl. Phys. **80**, 594 (1996)
26. I. Vurgaftman, J.R. Meyer, Band parameters for nitrogen-containing semiconductors. J. Appl. Phys. **94**, 3675 (2003)
27. H. Kawanishi, E. Niikura, M. Yamamoto, S. Takeda, Experimental energy difference between heavy- or light-hole valence band and crystal-field split-off-hole valence band in $Al_xGa_{1-x}N$. Appl. Phys. Lett. **89**, 251107 (2006)
28. K. Kamiya, Y. Ebihara, M. Kasu, K. Shiraishi, Efficient structure for deep-ultraviolet light-emitting diodes with high emission efficiency: a first-principles study of AlN/GaN superlattice. Jpn. J. Appl. Phys. **51**, 02BJ11 (2012)
29. K. Kamiya, Y. Ebihara, K. Shiraishi, M. Kasu, Structural design of AlN/GaN superlattices for deep-ultraviolet light-emitting diodes with high emission efficiency. Appl. Phys. Lett. **99**, 151108 (2011)

Chapter 10
Novel Behaviors Related to III-Nitride Thin Film Growth

Toru Akiyama

In Chaps. 4 and 7, we have discussed an ab initio-based chemical potential approach that incorporates the free energy of gas phase, to determine surface structures and growth kinetics on III-nitride surfaces. It has been revealed that this theoretical approach is useful to analyze the influence of temperatures and gas pressure related to the epitaxial growth of III-nitride compounds. By the application of this method, novel behavior during epitaxial growth of III-nitride compounds can be clarified. In this chapter, we discuss the feasibility of theoretical approach to various phenomena during epitaxial growth of III-nitride compounds. Surface phase diagram calculations as functions of temperature and gas pressure are performed to clarify the roles of Mg and C incorporation into GaN surfaces during growth. Furthermore, kinetic processes of initial nitridation of Al_2O_3 substrate are investigated using kinetic Monte Carlo (kMC) simulations (See, Sect. 2.3) on the basis of the results of adsorption, desorption, and migration energies obtained by ab initio calculations. Several perspectives for more realistic simulations for novel behaviors related to III-nitride thin film growth are also discussed.

10.1   Structure and Electronic States of Mg Incorporated InN Surfaces

Owing to the discovery of narrow energy gap in InN, [1, 2] InN has attracted much attention as a promising material for applications such as high-efficiency solar cells, light-emitting diodes, and high-frequency transistors. To realize these devices, the

T. Akiyama (✉)
Department of Physics Engineering, Mie University, Tsu, Japan
e-mail: akiyama@phen.mie-u.ac.jp

© Springer International Publishing AG, part of Springer Nature 2018   185
T. Matsuoka and Y. Kangawa (eds.), *Epitaxial Growth of III-Nitride Compounds*,
Springer Series in Materials Science 269,
https://doi.org/10.1007/978-3-319-76641-6_10

ability to fabricate both p- and n-type InN through doping remains a key challenge. Especially, p-type doping in InN has been very difficult to achieve due to its propensity for n-type carrier formation. Recently, buried p-type conductivity has been confirmed on In-polar samples which are grown along [0001] orientation. This indicates the realization of p-type InN by Mg doping [3–6]. Furthermore, the reduction in electron concentration in the electron accumulation layer of the surface was observed by Hall measurements, suggesting carrier compensation due to p-type doping [5, 6]. Since these experimental results reveal donor-acceptor interactions on polar InN surfaces, they raise interesting issues such as the stability of Mg at the surface and its interaction with the electron accumulation layer. Although the stabilization of donors and acceptors on by the compensation mechanism has been envisioned using the highly precise full potential linearized augmented plane-wave method, [7] the formation of Mg-incorporated surfaces considering the growth condition and its effects on the electronic properties still remain unresolved problems.

The growth and p-type doping along nonpolar $(1\bar{1}00)$ and $(11\bar{2}0)$ orientations in InN is also important issue. These nonpolar planes are attractive because of the absence of polarization fields, [8, 9] whereas the growth of InN epitaxial films along polar orientations such as [0001] and $[000\bar{1}]$ may result in large polarization fields along the growth direction reducing the radiative efficiency [10]. X-ray photoemission spectroscopy (XPS) [11] and scanning photoelectron microscopy and spectroscopy [12] have observed electron accumulation on $(1\bar{1}00)$ planes, and density functional calculations for the reconstructions on nonpolar InN$(1\bar{1}00)$ and $(11\bar{2}0)$ surfaces have clarified that the surfaces with In adlayers are favorable under In-rich conditions [13, 14]. It is thus of importance to clarifying the stability of Mg-incorporated surfaces taking account the doping conditions and the compensation mechanisms due to Mg doping in nonpolar orientations.

Figure 10.1 shows the calculated formation energies of Mg-incorporated surfaces on InN(0001) and $(000\bar{1})$ surfaces as a function of In chemical potential using the slab models with 2×2 and $\sqrt{3} \times \sqrt{3}$ periodicity along with those for bare surfaces. The figures of formation energy include only those reconstructions that are found to be energetically favorable in some part of the phase space spanned by the chemical potentials. Here, the value of μ_{Mg} is estimated using (2.63) at Mg pressure of 1.0×10^{-7} Torr for $T = 725$ and 825 K for InN(0001) and $(000\bar{1})$ surfaces, respectively. In the absence of Mg, the surfaces with an In adatom (In$_{ad}$) and a laterally contracted In bilayer (In$_{bilayer}$) are stabilized for moderately and extreme In-rich conditions on the (0001) surface, respectively, while the surface with an In adlayer (In$_{adlayer}$) is favorable over the entire In range on the $(000\bar{1})$ surface [13, 15]. In addition to these reconstructions, the reconstructions with In bilayer and In trilayer proposed by the XPS [16] for InN(0001) and $(000\bar{1})$ surfaces, respectively, are stabilized only under extremely In-rich conditions satisfying $\mu_{In} \geq 0.1$ eV.

For the (0001) orientation, the 2×2 surface with three substitutional Mg atoms at the topmost In sites, Mg$_{In}$(3/4) is found to be favorable over a wide range of In among various surface reconstructions including bare surfaces, as shown in Fig. 10.1a. Besides, the surface with two substitutional Mg atoms, Mg$_{In}$(1/2), is

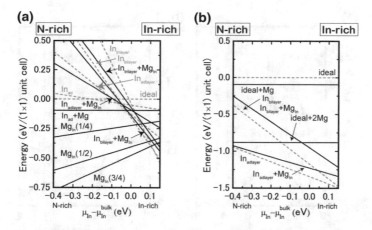

Fig. 10.1 Calculated formation energies of Mg-incorporated **a** InN(0001) and **b** InN(000$\bar{1}$) surfaces (solid lines) using (4.1) as a function of μ_{In}, along with those for bare surfaces (dashed lines). The value of μ_{Mg} are $E_{Mg} - 2.5$ and $E_{Mg} - 2.7$ eV, which correspond to Mg-rich conditions $(p_{Mg} = 1.0 \times 10^{-7}$ Torr$)$ at $T = 725$ and 825 K for InN(0001) and (000$\bar{1}$) surfaces, respectively. (Reproduced with permission from Akiyama et al. [78]. Copyright (2009) by American Physical Society.)

stabilized for low Mg pressures. This implies that Mg atoms can be easily incorporated on the surface and the fraction of substitutional Mg atoms at the topmost In sites increases with higher Mg pressure, qualitatively consistent with the high Mg concentration obtained from secondary-ion-mass spectroscopy (SIMS) [6]. The atomic configuration of $Mg_{In}(3/4)$ is identical to the most stable one in Mg incorporated GaN(0001) surfaces, [17, 18] indicating that its stabilization can be interpreted in terms of the electron counting (EC) rule. [19] From total-energy differences among various structures, the energy deficit due to excess electrons on In dangling bond [20] is estimated to be 0.65 eV/electron and the bond energy of In–N is lower than that of Mg–N by −0.1 eV. These results indicate that excess electrons on In-dangling bonds crucially affect the structural stability of (0001) surface.

The calculated results for InN(000$\bar{1}$) surface are different from those on InN (0001) surface. The 2×2 surface with a substitutional Mg atom at the topmost In layer of In adlayer ($In_{adlayer} + Mg_{In}$) is the most stable among Mg-incorporated surfaces over the entire μ_{In} range, consistent with x-ray absorption fine-structure measurements [21]. However, its formation energy is higher than that of In adlayer, especially for In-rich conditions (Fig. 10.1b). The formation-energy difference between these structures is small under N-rich conditions. Therefore, the Mg-incorporated InN(000$\bar{1}$) surfaces are always metastable under In-rich conditions and appear occasionally under N-rich conditions [22]. The stability of the $In_{adlayer}$ can be qualitatively understood by the preference of metallic In-In bonds compared to In-Mg bonds in the adlayer.

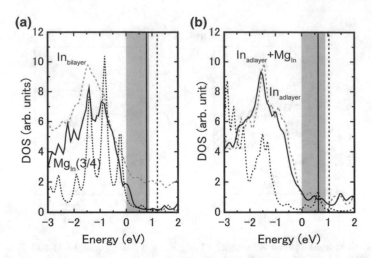

Fig. 10.2 Calculated DOS of **a** laterally contracted In bilayer structure $\left(\text{In}_{\text{bilayer}}\right)$ and the 2×2 surface with three substitutional Mg atoms, $\text{Mg}_{\text{In}}(3/4)$, by Mg incorporation on InN(0001) surface, and for **b** In adlayer reconstruction $\left(\text{In}_{\text{adlayer}}\right)$ and In adlayer structure with substitutional Mg $\left(\text{In}_{\text{adlayer}} + \text{Mg}_{\text{In}}\right)$ on InN(0001) surface. The local DOS of N atom near the surface for $\text{Mg}_{\text{In}}(3/4)$ and $\text{In}_{\text{adlayer}} + \text{Mg}_{\text{In}}$ is also shown by dotted lines. The energy gap obtained from bulk valence band maximum (VBM) and conduction band minimum in the band structure of Mg-incorporated InN surfaces is shown by shaded region. The zero of energy is set at the VBM. The vertical solid and dashed lines denote the Fermi energies of Mg-incorporated and bare surfaces, respectively. (Reproduced with permission from Akiyama et al. [78]. Copyright (2009) by American Physical Society.)

Figure 10.2 shows the calculated density of states (DOS) of Mg-incorporated surfaces, along with those of bare surfaces. The absence of electronic states around the Fermi energy is found in the DOS of $\text{Mg}_{\text{In}}(3/4)$ (solid line in Fig. 10.2a), exhibiting semiconducting character satisfying the EC rule [19]. Comparison with the DOS of Inbilayer dashed line in Fig. 10.2a also clarifies that the electronic states, which could be responsible for the electron accumulation, are reduced by Mg incorporation, leading to the reduction in electron concentration reported in the experiments [3, 4, 6]. In contrast, around the Fermi energy the DOS of $\text{In}_{\text{adlayer}} + \text{Mg}_{\text{In}}$ is similar to that of $\text{In}_{\text{adlayer}}$, as shown in Fig. 10.2b. This is because the states near the Fermi energy correspond to metallic In-In bonds in the In adlayer and there are little changes even after substitution by one Mg atom. It should be noted that the local DOS for N atoms near Mg at the surface for MgIn(3/4) and for $\text{In}_{\text{adlayer}} + \text{Mg}_{\text{In}}$ shown in Fig. 10.2a, b, respectively, are virtually occupied by electrons: Mg acceptors at the surface are compensated by electrons from the surfaces, supporting the experimental data [3, 4, 6].

Figure 10.3 shows the calculated formation energies of $\text{InN}(1\bar{1}00)$ and $(11\bar{2}0)$ surfaces as a function of In chemical potential using slab models with 1×2 and 1×1 periodicity for $\text{InN}(1\bar{1}00)$ and $(11\bar{2}0)$ surfaces, respectively. In the absence of Mg, the surfaces with a laterally contracted In bilayer, $\text{In}_{\text{bilayer}}$, and with an In

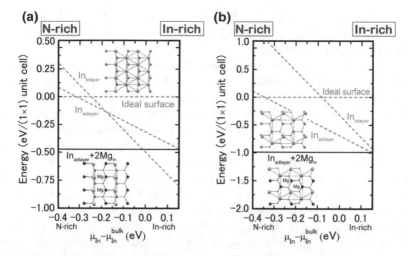

Fig. 10.3 Calculated formation energies of Mg-incorporated **a** InN($1\bar{1}00$) and **b** InN($11\bar{2}0$) surfaces (solid lines) using (4.1) as a function of μ_{In}, along with those for bare surfaces (dashed lines). The value of μ_{Mg} is $E_{\mathrm{Mg}} - 2.5\,\mathrm{eV}$, which corresponds to Mg-rich conditions $\left(p_{\mathrm{Mg}} = 1.0 \times 10^{-7}\,\mathrm{Torr}\right)$ at $T = 725\,\mathrm{K}$. Schematic top views of surfaces are also shown. Large and small gray circles represent In and N atoms, respectively. Black circles denote Mg atoms and light-gray circles top layer In atoms. The unit cell used in this study is shown by dashed rectangle. (Reproduced with permission from Akiyama et al. [78]. Copyright (2009) by American Physical Society.)

adlayer, In$_{\mathrm{adlayer}}$, shown in Fig. 10.3a, b are favorable for the ($1\bar{1}00$) and ($11\bar{2}0$) surfaces, respectively. For InN($11\bar{2}0$) surface, the calculated formation energy of In$_{\mathrm{bilayer}}$ is higher than that of In$_{\mathrm{adlayer}}$. Since the surface area per topmost atom on InN($11\bar{2}0$) surface is small, the distances between In adatoms in In$_{\mathrm{bilayer}}$ on the ($11\bar{2}0$) surface 3.02 Å become smaller than the In-In bond length of 3.28 Å in bulk In. The energy profit by forming In bilayer on the ($11\bar{2}0$) surface seems to be smaller than that on the ($1\bar{1}00$) surface due to stronger Coulomb repulsion.

It should be noted that there is a stable Mg-incorporated surfaces whose formation energies are lower than those for bare surfaces over the entire In range, in contrast to the results of nonpolar GaN($1\bar{1}00$) surfaces, [23] The structural features of top In and N atoms in these Mg-incorporated reconstructions are similar to those on InN(0001) and InN(000$\bar{1}$) surfaces, respectively: all In adatoms at the top layer of the ideal surface are substituted by Mg atoms and topmost N atoms are attached to In adatoms forming an In adlayer with In-In bonds as shown in Fig. 10.3a, b (In$_{\mathrm{adlayer}}$+2Mg$_{\mathrm{In}}$, hereafter). The stabilization of these structures results from the formation of covalent In-In and In-N bonds: Compared to the ideal surface, the energy gain of Mg incorporation is $\sim 0.5\,\mathrm{eV}$ per Mg atom. This comes from the energies of In-N bond (2.0 eV) and covalent In-In bonds (1.5 eV), the energy difference between In-N and Mg-N bonds ($\sim -0.1\,\mathrm{eV}$), and the chemical-potential shift due to the gas-phase Mg (about $-2.5\,\mathrm{eV}$). Since the energy gain for Mg incorporation in InN($11\bar{2}0$) surface (0.16–0.60 eV per Mg atom) is the largest over the wide range of In chemical potential, it is expected that the ($11\bar{2}0$) plane is most preferred for Mg incorporation.

The reduction in surface carrier concentrations and the compensation of Mg acceptors on nonpolar planes can also be deduced from the DOS in Mg-incorporated InN($1\bar{1}00$) and ($11\bar{2}0$) surfaces. Figure 10.4 illustrates the DOS of $In_{adlayer}+2Mg_{In}$, along with those of stable bare surfaces such as $In_{bilayer}$ and $In_{adlayer}$. The calculated DOS demonstrates that the electronic states around the Fermi energy in bare surfaces (dashed line in Fig. 10.4) are reduced in $In_{adlayer}+2Mg_{In}$ (solid line in Fig. 10.4), suggesting the reduction in electron concentration by Mg doping. The analysis of wave functions clarifies that the reduction in electronic states comes from the electron transfer from the In adlayer to Mg forming the bonding states between Mg and N atoms and the formation of covalent In-In bonds along the [$11\bar{2}0$] and [0001]directions for InN($1\bar{1}00$) and ($11\bar{2}0$) surfaces, respectively. Therefore, $In_{adlayer}+2Mg_{In}$ is stabilized and exhibits a semiconducting nature. It is also found that the local DOS for N atoms near the surface for $In_{adlayer}+2Mg_{In}$ (dotted lines in Fig. 10.4) are occupied by electrons. Similar to the case of Mg-incorporated polar surfaces, Mg acceptors located at the nonpolar surfaces are expected to be compensated by electrons from the surfaces.

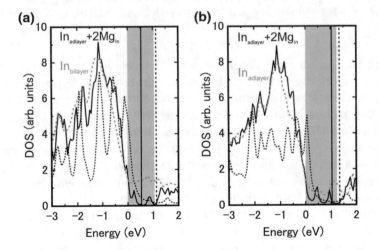

Fig. 10.4 Calculated DOS of **a** laterally contracted In-bilayer structure (Inbilayer) and In-monolayer surface with two substitutional Mg atoms $\left(In_{adlayer}+2Mg_{In}\right)$ by Mg incorporation on InN($1\bar{1}00$) surfaces, and for **b** InN($11\bar{2}0$) surface with In adlayer $\left(In_{adlayer}\right)$ and In adlayer surface with two substitutional Mg atoms $\left(In_{adlayer}+2Mg_{In}\right)$. Note that the gap energies obtained by the calculations of InN($1\bar{1}00$) and ($11\bar{2}0$) surfaces are larger than that in the bulk case by 0.4 and 0.5 eV, respectively. The same notation as in Fig. 10.2. (Reproduced with permission from Akiyama et al. [78]. Copyright (2009) by American Physical Society.)

10.2   Magnesium Incorporation on Semipolar GaN($1\bar{1}01$) Surfaces

The discovery of p-type conductivity in GaN using Mg doping leads to the widespread development of GaN-based optoelectronic device, such as light-emitting diodes, laser diodes, and photovoltaic cells [24–27]. However, GaN-based heterostructures are usually grown along the polar [0001] direction which has large polarization related electric field inside the multiquantum wells [28]. The internal electric fields in the spontaneous and piezoelectric polarization give rise to the quantum-confined Stark effect and reduce the radiative efficiency within the quantum wells. In order to reduce the effects of internal electric fields, there is an increasing interest in the crystal growth and device fabrication on semipolar orientations such as ($11\bar{2}2$), ($1\bar{1}03$), and ($1\bar{1}01$), due to their reduced and even negligible electric field [29–38]. It has been recently found that the incorporation of Mg is more efficient on a GaN($1\bar{1}01$) surface than on polar (0001) surfaces in addition to reduction of the internal electric fields: Mg concentrations on GaN($1\bar{1}01$) surface measured by SIMS are higher than that on polar (0001) surface [36, 38]. However, the ideal GaN($1\bar{1}01$) surface is N-terminated surface such as GaN($000\bar{1}$) surface, in which the Mg doping efficiency is rather poor [39]. Therefore, the origins for high Mg concentrations on N-terminated GaN($1\bar{1}01$) surface cannot be explained by the analogy with GaN($000\bar{1}$) surface. To clarify the microscopic origin for high Mg concentrations on semipolar ($1\bar{1}01$) orientation, investigations for Mg-incorporation behavior as well as the reconstructions on the ($1\bar{1}01$) surfaces from theoretical viewpoints are necessary.

From theoretical viewpoints, the stability of Mg on GaN surfaces has been intensively investigated to address many of the issues raised by the experimental results. To explain the narrow window for smooth growth of GaN due to Mg on GaN(0001) surface, [40] the relative stability of possible Mg-rich reconstructions has been determined with respect to those of the clean surface, and the surface structures comprising 1/2–3/4 monolayer of Mg substituting for Ga in the top layer have been proposed in very Mg-rich conditions [17]. Sun et al. [18] have investigated the energetics of Mg adsorption and incorporation at GaN(0001) and ($000\bar{1}$) surfaces under a wide variety of conditions. They have clarified that the Mg incorporation is easier at the Ga-polar surface, but high Mg coverages tend to locally change the polarity from Ga to N polar. A thermodynamic approach with chemical potentials appropriate for realistic growth conditions has revealed that hydrogen stabilizes Mg-rich surface reconstructions for both GaN(0001) and ($1\bar{1}00$) surfaces [23]. To explain high hole concentrations in Mg-doped semipolar GaN($1\bar{1}0\bar{1}$) surface, [33] the stability of Mg-incorporated GaN($1\bar{1}0\bar{1}$) surface has also been examined [41]. Furthermore, the effects of hydrogen on the incorporation of Mg on semipolar GaN($1\bar{1}01$) surface have been clarified on the basis of surface phase diagrams [42].

Figure 10.5 shows the stable structures on GaN($1\bar{1}01$) surface including Mg as functions of μ_{Ga} and μ_{Mg} using (4.1) under low H_2 pressures, along with that on GaN(0001) surface for comparison. The boundary lines separating different regions correspond to chemical potentials for which two structures have the same formation energy. Here, only one Mg atom in the unit cell is assumed because the partial pressure of Mg during doping should be much lower than that of Ga: The chemical potential of Mg is expected to vary much lower than the Mg-rich limit for Mg_3N_2 precipitation (dashed lines in Fig. 10.5). The Mg-rich limit is calculated by assuming the surface in equilibrium with bulk Mg_3N_2 expressed as

$$3\mu_{Mg} + 2\left(\mu_{GaN}^{bulk} - \mu_{Ga}\right) = \mu_{Mg_3N_2}^{bulk}, \tag{10.1}$$

where μ_{GaN}^{bulk} and $\mu_{Mg_3N_2}^{bulk}$ are chemical potentials of bulk phase GaN and Mg_3N_2. By taking account of different number of Mg atoms, similar formation energies reported in the previous calculations for GaN(0001) surface [18, 23] are obtained. These figures clearly demonstrate that Mg atoms can be incorporated when μ_{Mg} is higher than -1.6 eV, depending on μ_{Ga}. For GaN($1\bar{1}01$) surface, as shown in Fig. 10.5a, the surface with a substitutional Mg at one of Ga–Ga dimers (Ga–Ga dimers+Mg_{Ga} shown in Fig. 10.6a) can be stabilized for $\mu_{Ga} = -0.58$ eV while the surface with a Ga monolayer including Mg (Ga monlayer + Mg_{Ga} shown in Fig. 10.6b) is favorable under Ga-rich conditions ($\mu_{Ga} = -0.58$ eV). In the case of GaN(0001) surface, the ideal surface with a substitutional Mg (ideal surface + Mg_{Ga} in Fig. 10.5b) is favorable over the wide range of μ_{Ga} if μ_{Mg} is lower than -0.5 eV.

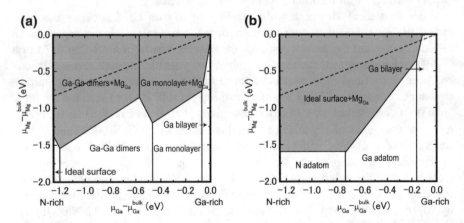

Fig. 10.5 Stable structures on Mg-incorporated **a** GaN($1\bar{1}01$) and **b** GaN(0001) surfaces as functions of μ_{Ga} and μ_{Mg} for low H_2 pressures ($\mu_H = (1/2)E_{H_2} - 2.3$ eV, where E_{H_2} is the energy of H_2 molecule). The stable regions of Mg-incorporated surfaces, such as Ga-Ga dimer+Mg_{Ga} and Ga monlayer+Mg_{Ga} are emphasized by shaded area. Geometries of Mg-incorporated GaN($1\bar{1}01$) surfaces are shown in Fig. 10.6. Dashed lines denote the Mg-rich limit conditions expressed by (10.1). μ_{Ga}^{bulk} and μ_{Mg}^{bulk} are the energies of bulk Ga and Mg, respectively. (Reproduced with permission from Akiyama et al. [42]. Copyright (2010) by American Physical Society.)

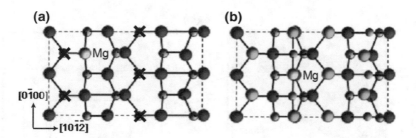

Fig. 10.6 Schematic top views of Mg-incorporated GaN($1\bar{1}01$) surfaces with a substitutional Mg in **a** Ga-Ga dimers and **b** Ga monolayer, which are stabilized under N-rich and Ga-rich conditions, respectively. Large and small gray circles represent Ga and N atoms, respectively. The Mg atoms are represented by black circles. The positions of H atoms in the H terminated surface (4N–H +Mg$_{\text{Ga}}$ in Fig. 10.5) are marked by crosses in the Mg-incorporated surface with Ga-Ga dimers. (Reproduced with permission from Akiyama et al. [42]. Copyright (2010) by American Physical Society.)

In addition, the stable temperature range of the Mg-incorporated surface under the N-rich limit can be estimated by comparing the value of μ_{Mg} at the boundary line in Fig. 10.5 with the gas-phase chemical potential of Mg atom using (2.63). When the pressure of Mg is lower than 1.0×10^{-2} Torr, the transition temperatures are found to be 1010–1400 K for GaN($1\bar{1}01$) surface and 1050–1430 K for GaN (0001) surface: For low H$_2$ pressures, Mg atoms can be incorporated regardless of the growth orientation, indicating that high Mg concentrations can be obtained in both ($1\bar{1}01$) and (0001) orientations in the case of molecular beam epitaxy (MBE) growth. However, this results differ from the SIMS measurements of GaN fabricated by the MOVPE growth, in which the concentrations of Mg on GaN($1\bar{1}01$) surface is higher than that on GaN(0001) surface [36, 38]. Due to the presence of hydrogen, the incorporation behavior of Mg in the case of metal-organic vapor-phase epitaxy (MOVPE) growth is different from that in the MBE growth, as explained below.

Figure 10.7 depicts the stable structures on GaN($1\bar{1}01$) surface including Mg atoms as functions of μ_{Ga} and μ_{Mg} using (4.1) under high H$_2$ pressures, along with that on GaN(0001) surface. These diagrams demonstrate that Mg atoms can be incorporated, but there is remarkable orientation dependence in the stable regions of the Mg-incorporated surfaces. As shown in Fig. 10.7a, for GaN($1\bar{1}01$) surface the H-terminated surface with substitutional Mg (4N–H + Mg$_{\text{Ga}}$) are stabilized over the wide chemical potential range of Ga. Although the stable Mg-incorporated structure for high H$_2$ pressures is different from those for low H$_2$ pressures, trends in the stable region of Mg-incorporated surfaces among the phase space spanned by the chemical potentials are similar with each other. In contrast, for GaN(0001) surface the stable region of Mg- incorporated surfaces is drastically reduced by the presence of hydrogen. For $\mu_{\text{Mg}} \leq -0.91$ eV, the H-terminated surface with N adatom [N$_{\text{ad}}$–H + Ga–H in Fig. 10.7b] is stabilized over the wide range of μ_{Ga}. This is

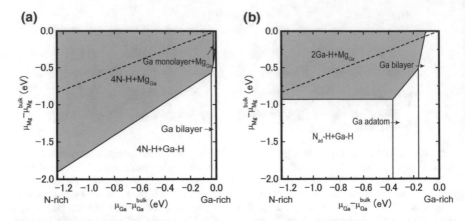

Fig. 10.7 Stable structures on Mg-incorporated **a** GaN($1\bar{1}01$) and **b** GaN(0001) surfaces as functions of μ_{Ga} and μ_{Mg} for high H_2 pressure ($\mu_H = (1/2)E_{H_2} - 1.05$ eV, where E_{H_2} is the energy of H_2 molecule). The notation is the same as in Fig. 10.5. (Reproduced with permission from Akiyama et al. [42]. Copyright (2010) by American Physical Society.)

because strong Ga–N and N–H bonds are formed in the N_{ad}–H + Ga–H, which simultaneously satisfies the EC rule. [19] Under N-rich conditions, the energy range of μ_{Mg} in which the Mg-incorporated surface is stabilized decreases due to the stabilization of the N_{ad}–H + Ga–H.

We can estimate surface phase diagrams for Mg incorporation as functions of temperatures and pressure for phase transition between Mg-incorporated and Mg-free surfaces under N-rich limit, and discuss the orientation dependence in the stability of Mg-incorporated surfaces under high H_2 pressures more quantitatively. Figure 10.8 shows the temperatures for phase transition on GaN($1\bar{1}01$) and GaN (0001) surfaces at $p_{H_2} = 76$ Torr. For GaN($1\bar{1}01$) surface, the temperatures at $p_{H_2} = 76$ Torr (1090–1530 K) are almost similar to those at $p_{H_2} = 1.0 \times 10^{-8}$ Torr (1010–1400 K), as shown in Fig. 10.8a. On the contrary, as shown in Fig. 10.8b, the transition temperatures in GaN(0001) surface at $p_{H_2} = 76$ Torr are ranging 930–1310 K, which are remarkably lower than those at $p_{H_2} = 1.0 \times 10^{-8}$ Torr (1050–1430 K). The pressure dependence in the transition temperatures for GaN(0001) surface originates from the difference in the boundary line between the surfaces with and without Mg, leading to the orientation dependence in the transition temperature. The lower temperatures for the phase transition on GaN(0001) thus suggest that under the MOPVE growth around 1300 K the incorporation of Mg atoms on semipolar ($1\bar{1}01$) orientation is rather efficient than that on the polar (0001) orientation. Although the kinetics such as adsorption and desorption behavior of Mg during the growth processes should be verified, this efficient Mg incorporation results in high Mg concentrations on GaN($1\bar{1}01$) surface, qualitatively consistent with the experimental results by the SIMS [36, 38].

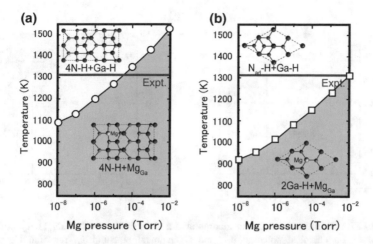

Fig. 10.8 Surface phase diagram of Mg-incorporated and Mg-free surfaces in the N-rich limit ($\mu_{Ga} - \mu_{Ga}^{bulk} = -1.24$ eV, where μ_{Ga}^{bulk} is the energy of bulk Ga) as a function Mg pressure on **a** GaN($1\bar{1}01$) and **b** GaN(0001) surfaces. The Mg-incorporated surfaces are stabilized in the regions emphasized by shaded area. Schematic views of surface structures are also shown. Large and small gray circles rep- resent Ga and N atoms, respectively. The H and the Mg atoms are represented by tiny and larger black circles, respectively. The growth temperature of GaN by the MOVPE in [35, 38] is attached by dashed lines. (Reproduced with permission from Akiyama et al. [42]. Copyright (2010) by American Physical Society.)

10.3 Carbon Incorporation on Semipolar GaN($1\bar{1}01$) Surfaces

It is well known that the most widely used dopant for p-type GaN is Mg. Carbon also acts as a p-type dopant if it is incorporated on the nitrogen lattice site [43]. However, successful p-type doping has never been reported by carbon doping on conventional GaN(0001) substrate. Previous experiments have reported that the carbon doping on this orientation results in the formation of deep levels [44]. There has been an increasing interest in the crystal growth and device fabrication on semipolar orientations [29–38, 45] owing to their reduced and even negligible electric field. In addition to eliminating the effects of polarization induced electric fields, p-type conductivity is successfully obtained by the carbon doping on semipolar GaN($1\bar{1}01$) surface. [34, 37] These experimental findings thus suggest that the difference in doping behavior between (0001) and ($1\bar{1}01$) orientations can be attributed to the polarity of these sur- faces: the ideal GaN(0001) and ($1\bar{1}01$) surfaces are basically terminated by Ga and N faces, respectively, which results in the difference in the substitution efficiency of carbon on the nitrogen lattice sites.

The difference in carbon incorporation behavior depending on the surface orientation is clarified by examining the stability and structure of carbon incorporated surfaces [42]. Figure 10.9 depicts the diagrams of stable structures on GaN(0001)

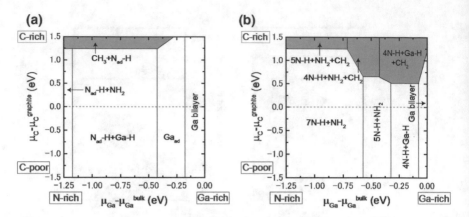

Fig. 10.9 Stable structures of carbon incorporated **a** GaN(0001) and **b** GaN($1\bar{1}01$) surfaces as functions of Ga chemical potential μ_{Ga} and C chemical potential μ_C for high H_2 pressure conditions ($\mu_H = (1/2)E_{H_2} - 1.05$ eV, where E_{H_2} is the energy of H_2 molecule). The stable regions of carbon incorporated surfaces are emphasized by shadedareas. Geometries of GaN($1\bar{1}01$) surfaces are shown in Fig. 10.10. μ_{Ga}^{bulk} and $\mu_C^{graphite}$ are the energies of bulk Ga and graphite, respectively. (Reproduced with permission from Akiyama et al. [79]. Copyright (2011) by the Japan Society of Applied Physics.)

and ($1\bar{1}01$) surfaces including carbon as functions of μ_{Ga} and μ_C. Here, the value of hydrogen chemical potential $\mu_H = (1/2)E_{H_2} - 1.05$ eV (E_{H_2} is the total energy of H_2 molecule) obtained by the ideal diatomic gas model is used to calculate the surface formation energy in (4.1). The value of -1.05 eV corresponds to the MOVPE growth. These diagrams demonstrate that carbon-free surfaces, which are described in Sect. 4.2, are stabilized over the wide range of μ_{Ga} and μ_C. However, a carbon atom can be incorporated under C-rich condition both for GaN(0001) and ($1\bar{1}01$) surfaces. Besides, there is an orientation dependence in the stabilization of carbon incorporated structures. In the case of the GaN(0001) surface, as shown in Fig. 10.9a, the surface with CH_3 and H-terminated N adatoms ($CH_3 + N_{ad}$–H) is stabilized for $\mu_C - \mu_C^{graphite} \geq 1.24$ eV ($\mu_C^{graphite}$ is the chemical potential of graphite) under N-rich and moderate Ga-rich conditions. This structure corresponds to the substitution of NH_2 by CH_3 in the hydrogen terminated surface with NH_2 and N adatoms (N_{ad}–H + NH_2), implying that under C-rich conditions C atoms can be preferentially adsorbed on the N lattice site on the GaN(0001) surface. In contrast, the chemical potential range for the stabilization of the carbon incorporated GaN($1\bar{1}01$) surface shown in Fig. 10.9b is larger than that on the GaN(0001) surface. The surface with CH_2 at the Ga lattice site ($5N$–H + NH_2 + CH_2) shown in Fig. 10.10b is stabilized for $\mu_C - \mu_C^{graphite} \geq 1.24$ eV under N-rich conditions, and those with CH_2 at the N lattice site ($4N$–H + NH_2 + CH_2 and $4N$–H + Ga–H + CH_2) shown in Fig. 10.10c, d are stable for $\mu_C - \mu_C^{graphite} = 0.5$ eV under Ga-rich conditions. Therefore, p-type conductivity on the GaN($1\bar{1}01$) surface can be realized via the formation of $5N$–H + NH_2 + CH_2 ($4N$–H + NH_2 + CH_2 and $4N$–H + Ga–H

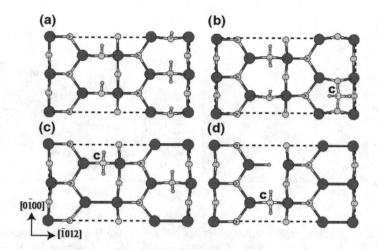

Fig. 10.10 Schematic top views of hydrogen terminated GaN($1\bar{1}01$) surfaces with **a** three N–H bonds and NH$_2$ (7N–H + NH$_2$ in Fig. 10.9), **b** CH$_2$ at the Ga site (5N–H + NH$_2$ + CH$_2$), **c** NH$_2$ and CH$_2$ at the N site (4N–H + NH$_2$ + CH$_2$), and **d** Ga–H and CH$_2$ at the N site (4N–H + Ga–H + CH$_2$). Large, small, and tiny circles represent Ga, N, and H atoms, respectively. The C atoms represented by shaded circles are labeled. The unit cell used in this study is shown by a dashed rectangle. (Reproduced with permission from Akiyama et al. [79]. Copyright (2011) by the Japan Society of Applied Physics.)

+ CH$_2$) under N-rich (Ga-rich) conditions. The orientation dependence in the stability of carbon incorporated surfaces is due to the formation of Ga–C (N–C) bonds. The energy profits caused by two Ga–C and N–C bonds (5.8 and 5.0 eV, respectively) in the carbon incorporated GaN($1\bar{1}01$) surface are larger than that of a single Ga–C bond (3.1 eV) on the GaN(0001) surface. The contribution of excess or deficit electrons on the surface dangling bonds is negligible because the carbon incorporated surfaces obtained in this study satisfy the EC rule [19]. It should be noted that all the carbon incorporated structures are thermodynamically unstable against the formation of graphite. However, the conditions of μ_C higher than that of graphite would be realized during the growth due to the higher μ_C in C$_2$H$_4$ and CCl$_4$, which are used as the source gases [34, 37]. It seems that the high reaction energy of graphite from these source gases prevents the graphite formation during the growth of GaN.

Furthermore, by using surface phase diagrams, we can estimate the temperature for carbon desorption in the N-rich limit, and discuss the orientation dependence in the stability of carbon incorporated surfaces more quantitatively. Figure 10.11 shows the calculated surface phase diagrams for the desorption of carbon atoms on GaN(0001) and ($1\bar{1}01$) surfaces as functions of temperature and carbon pressure obtained by using (2.63). These surface diagrams imply that for both GaN(0001) and ($1\bar{1}01$) surfaces the incorporation of carbon during the growth is efficient for low temperatures. For the GaN(0001) surface, the temperatures for the desorption of C on CH$_3$ + N$_{ad}$–H range from 1550–2050 K depending on C pressure. In

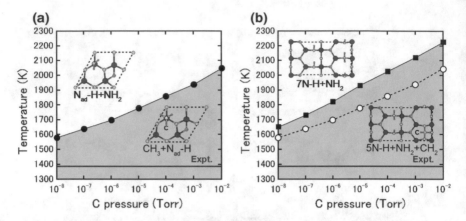

Fig. 10.11 Calculated surface phase diagrams of carbon incorporated **a** GaN(0001) and **b** GaN($1\bar{1}01$) surfaces as functions of temperatures and carbon pressure in the N-rich limit ($\mu_{Ga} - \mu_{Ga}^{bulk} = -1.24$ eV, where μ_{Ga}^{bulk} is the energy of bulk Ga). The carbon incorporated GaN surfaces are stabilized in the regions emphasized by shaded areas. Schematic views of surface structures are also shown. The growth temperature in [34, 37] is shown by dashed lines. The dashed line in **b** is the phase boundary between carbon free and incorporated surfaces on GaN (0001) surface for comparison. (Reproduced with permission from Akiyama et al. [79]. Copyright (2011) by the Japan Society of Applied Physics.)

contrast, the desorption temperatures of C on 5N–H + NH$_2$+CH$_2$ range from 1660 to 2220 K, which are higher than those on the GaN(0001) surface. This temperature difference leads to the orientation dependence in the stability under growth conditions. Considering the kinetics such as surface migration, more carbon atoms on the GaN(0001) surface compared to the GaN($1\bar{1}01$) surface could desorb during the growth processes. It is thus expected that the concentration of C atoms on the GaN($1\bar{1}01$) surface is higher than that on the GaN(0001) surface. This high carbon concentrations result in the realization of p-type doping only on the GaN($1\bar{1}01$) surface. It should be noted that the most stable adsorption site under N-rich condition [5N–H + NH$_2$ + CH$_2$ in Fig. 10.10b] is located at the Ga lattice site. If this C atom is stably located at the Ga lattice site during the growth processes, p-type conductivity cannot be explained by this structure. To find the percentage of carbon that can be ionized and release holes on the GaN($1\bar{1}01$) surface, detailed adsorption and desorption behaviors of carbon atoms during the growth processes should be clarified.

10.4 Stability of Nitrogen Incorporated Al$_2$O$_3$ Surfaces

For III-nitride compounds, sapphire (Al$_2$O$_3$) has been widely used as a substrate for their epitaxial growth. Furthermore, it is well known that the pregrowth treatment of the substrate such as the nitridation strongly affects the nucleation of materials, leading to the improvements of structural and optical properties in epitaxial GaN films [46–49]. The initial nitridation of sapphire surface is now one of essential techniques to fabricate high-quality nitrides in MOVPE, [46, 48, 50–56] MBE, [47, 49, 57–61] and hydride vapor phase epitaxy (HVPE) [62].

The nitridation of sapphire is usually performed by the exposure of sapphire substrate to NH_3 and nitrogen plasma in the MOVPE and MBE, respectively. The analysis of nitrided layers on $Al_2O_3(0001)$ surface in the MOVPE has reported the formation of amorphous aluminum oxynitride layer, [50, 51] while several studies have shown the formation of relaxed crystalline AlN [52, 53]. The overvations using TEM in the MBE on the (0001) surface have reported the formation of epitaxial AlN [57, 58]. The formation of AlN and NO molecules [59, 60] as well as oxynitride compound [61] has also been suggested in the MBE. More recently, the nitridation on the $(1\bar{1}02)$ orientation in the MOVPE has been found to improve the crystal quality and surface morphology of epitaxial GaN grown along nonpolar $(11\bar{2}0)$ direction [54, 55]. From these experimental results, there is a general consensus in which N atoms are incorporated into sapphire surface layers. Despite these experimental findings, there have been few theoretical investigations for the stability and structure of nitrogen incorporated Al_2O_3 surfaces [63].

10.4.1 $Al_2O_3(0001)$ Surface

Figure 10.12 depicts schematics of nitrogen incorporated $Al_2O_3(0001)$ surface. The ideal (0001) plane is composed of two Al layers and one O layer, and each Al (O layer in the 1×1 lateral unit is constructed of one Al atom (three O atoms). The 1×1 surface terminated by single Al layer shown in Fig. 10.12a (Al_2O_3–Al–I) is found to be the most stable surface structure. It is thus expected that Al_2O_3–Al–I corresponds to the 1×1 surface observed in the experiments for temperatures lower than 1400 K [64]. The stability of Al_2O_3–Al–I can be attributed to its stoichiometry where the stoichiometry of Al and O at the top three surface layers is identical to that in bulk Al_2O_3. There are no excess/deficit electrons at the surface and no electric dipole moment along the [0001] direction. The adsorption of an N atom is verified by placing one N adatom at various positions on the Al_2O_3–Al–I. There are two symmetrically inequivalent configurations for the surface with an N adatom. The most stable configuration is constructed by the N adatom located above the O atom in the second layer, so that there are three stable adsorption sites in the 1×1 unitcell (A1, A2, and A3 sites in Fig. 10.12a). The optimized geometry (Al_2O_3–N_{ad}) shown in Fig. 10.12b indicates the formation of covalent N–O bond. Another configuration is composed of the N adatom located above the outermost Al atom (B in Fig. 10.12a) but its energy is 2.03 eV higher than that of the stable configuration. In Al_2O_3–N_{ad}, there is a single covalent N–O bond between the N adatom and one of O atoms in the second layer along the [0001] direction. On the other hand, no covalent Al–N bond is recognized for the N adatom located above

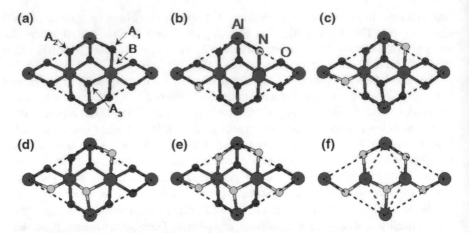

Fig. 10.12 Schematic top views of **a** the stoichiometric Al-terminated $Al_2O_3(0001)$surface (Al_2O_3–Al-I), and the (0001) surfaces with **b** one N adatom (Al_2O_3–N_{ad}), **c** one substitutional N atom, **d** two substitutional N atoms, **e** three substitutional N atoms (Al_2O_3–$3N_{sub}$), and (f) oxygen desorption (AlN-Al_2O_3). The 1×1 unit cell of $Al_2O_3(0001)$ and AlN(0001) surfaces are enclosed by dashed and dotted lines. The adsorption sites of N adatom on the Al_2O_3–Al-I are denoted by A_1, A_2, A_3, and B. (Reproduced with permission from Akiyama et al. [63]. Copyright (2012) by Elsevier.)

the outermost Al atom. This is because there are few electrons around the outermost Al atom. The adsorption of N atom with the covalent N–O bond in Al_2O_3–N_{ad} is exothermic, but the adsorption energy (-0.55 eV) is too small for the adsorption of N atoms under the pregrowth conditions. The nitrogen substitution for O atoms in the second layer shown in Fig. 10.12c–e can be considered as possible surface structures during the nitridation. Furthermore, the desorption of O atoms as well as the substitution is necessary to form AlN layer from Al_2O_3. The surface where three oxygen atoms located in the second layer of Al_2O_3–Al–I desorb and N atoms substitute for those in the fifth layer (AlN–Al_2O_3) shown in Fig. 10.12f is examined as a representative of surface structures with oxygen desorption.

Figure 10.13 shows stable surface structures on $Al_2O_3(0001)$ surface as functions of μ_O and μ_N. There are several stable structures appear depending on the pregrowth conditions. The stoichiometric Al_2O_3–Al–I is stabilized over the wide range of O-rich and N-poor conditions, even though Al_2O_3–N_{ad} is favorable in a very narrow chemical potential range ($\mu_N - \mu_{N_{atom}} \geq 0.55$, where $\mu_{N_{atom}}$ is the energy of N atom) eV. The surfaces with one and two substitutional N atoms shown in Fig. 10.12c, d, respectively, are always metastable over the entire chemical potential range. When μ_O satisfies $\left(\mu_O - \mu_{O_2}\right) - \left(\mu_N - \mu_{N_{atom}}\right) \leq -1.02$ eV (μ_{O_2} is the energy per atom of O_2 molecule) and $\mu_O - \mu_{O_2} \geq -2.86$ eV, three O sites (in the 1×1 unit cell) at the second layer are completely replaced by N atoms (Al_2O_3–$3N_{sub}$) as shown in Fig. 10.12e. Furthermore, AlN–Al_2O_3 is found to be stabilized under N-rich and O-poor conditions satisfying $2\left(\mu_O - \mu_{O_2}\right) - \left(\mu_N - \mu_{N_{atom}}\right) \leq -3.88$ eV and

Fig. 10.13 Stable structures
of nitrogen incorporated
$Al_2O_3(0001)$ surface as
functions of oxygen chemical
potential μ_O and nitrogen
chemical potential μ_N. The
pressure of O_2 molecule (N
atoms) at 1300 K is related to
μ_O (μ_N) using the ideal gas
model. (Reproduced with
permission from Akiyama
et al. [63]. Copyright (2012)
by Elsevier.)

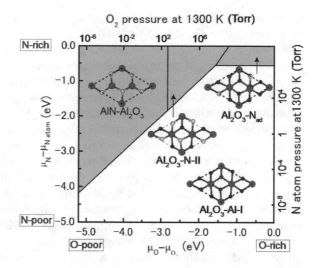

$\mu_O - \mu_{O_2} \leq -2.86$ eV. Since the value of $\mu_O - \mu_{O_2} = -2.86$ eV corresponds to the O_2 pressure of $\sim 10^2$ Torr at 1300 K, it is expected that the desorption of oxygen from $Al_2O_3(0001)$ surface occurs under the nitridation conditions. The oxygen desorption can be interpreted in terms of the decrease in energy by forming Al–N bonds instead of Al–O bonds.

Another important structural feature of $AlN–Al_2O_3$ is found in the remaining first and second layers. The positions of substitutional N atoms become quite different from those of O sites on $Al_2O_3(0001)$surface. Consequently, these layers consist of Al and N atoms each of which in-plane position is almost identical to each Al and N site on AlN(0001) surface. This leads to the formation of AlN layers whose in-plane alignment satisfies $AlN[1\bar{1}00]\|Al_2O_3[11\bar{2}0]$. The formation of AlN layers with this in-plane alignment also implies that the alignment between epitaxially grown nitrides on the AlN layer and $Al_2O_3(0001)$surface becomes $[1\bar{1}00]$ nitrides $\|[11\bar{2}0]$ sapphire. Therefore, the improvement of structural and optical properties of nitrides after the nitridation [46–49] can be attributed to AlN layers which are generated by the oxygen desorption. Indeed, the formation of AlN layers is consistent with the experiments, [52, 53, 57–60] and the calculated in-plane alignment agrees well with those obtained between GaN and sapphire substrate [65, 66].

10.4.2 $Al_2O_3(1\bar{1}02)$ Surface

Similar to $Al_2O_3(0001)$ surface, the adsorption of an N atom by placing one N atom at various points on the Al_2O_3–O–I shown Fig. 10.14a occurs. The $Al_2O_3(1\bar{1}02)$ surface is composed of two Al layers and three O layers each of which in the 1×1

lateral unitcell is constructed of two atoms, leading to the alternately stacked-layer unit expressed as O_2–Al_2–O_2–Al_2–O_2. The 1×1 surface terminated by single O layer shown in Fig. 10.14a (Al_2O_3–O–I) is the most stable irrespective of the chemical potentials of constituting elements. In this structure, the stoichiometry of Al and O at the top five surface layer is identical to that in bulk Al_2O_3. It is thus expected that this reconstruction corresponds to the 1×1 surface observed in the experiments for temperatures lower than 1700 K [67]. The most stable adsorption sites are located above the outermost O atoms (A_1 and A_2 in Fig. 10.14a) and the optimized geometry (Al_2O_3–N_{ad}) shown in Fig. 10.14b indicates the formation of a covalent N–O bond. The other adsorption sites are found be located above the O atom in the third layer (B_1 and B_2 in Fig. 10.14a). Although a covalent N–O bond is formed between the N adatom at B_1 (B_2) site and the O atom in the third layer, the energies of the surfaces with an N adatom at these sites are 0.16 eV higher than those at A_1 and A_2 sites. The adsorption of an N adatom at A_1 and A_2 sites is exothermic and its adsorption energy is -0.55 eV. The high adsorption energy, however, indicates that the surface with an N adatom is not stabilized under the pregrowth conditions. The nitrogen substitution for the outermost O atoms shown in Fig. 10.14c, d is thus taken into account as possible surface structures during the

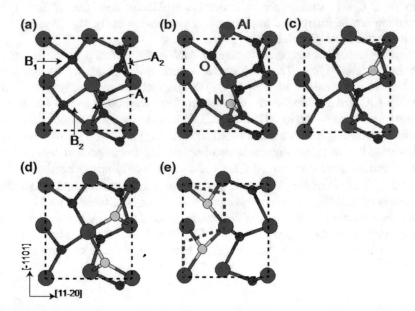

Fig. 10.14 Schematic top views of **a** the stoichiometric O-terminated $Al_2O_3(1\bar{1}02)$ surface (Al_2O_3–O–I), and the $(1\bar{1}02)$ surfaces with **b** one N adatom (Al_2O_3–N_{ad}), **c** one substitutional N atom, (d) two substitutional N atoms (Al_2O_3–$2N_{sub}$), and **e** oxygen desorption (AlN–Al_2O_3). The notation is the same as that in Fig. 10.12. The ideal positions of Al and N atoms at the top layer of $Al_2O_3(1\bar{1}02)$ surface is represented by dotted lines. The adsorption sites of N adatom on the Al_2O_3–O–I are denoted by A_1, A_2, B_1, and B_2. (Reproduced with permission from Akiyama et al. [63]. Copyright (2012) by Elsevier.)

Fig. 10.15 Stable structures of nitrogen incorporated $Al_2O_3(1\bar{1}02)$ surface as functions of oxygen chemical potential μ_O and nitrogen chemical potential μ_N. The notation is the same as that in Fig. 10.13. (Reproduced with permission from Akiyama et al. [63]. Copyright (2012) by Elsevier.)

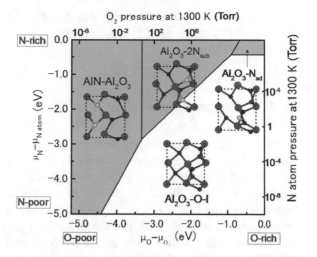

nitridation. To verify the oxygen desorption, the surface where oxygen atoms in the top layer desorb and N atoms substitute for those in the third layer (AlN–Al_2O_3) shown in Fig. 10.14e is also considered.

Figure 10.15 shows the stable surface structures of $Al_2O_3(1\bar{1}02)$ surface as functions of μ_O and μ_N, representing the stabilization of nitrogen incorporated surfaces. The stoichiometric Al_2O_3–O–I is the most stable over the wide range of O-rich and N-poor conditions. However, two oxygen sites (in the 1×1 unit cell) at the top layer of the surface are replaced by N atoms (Al_2O_3–$2N_{sub}$ shown in Fig. 10.14d) under moderate and O-poor conditions satisfying $(\mu_O - \mu_{O_2}) - (\mu_N - \mu_{N_{atom}}) \leq -0.50$ eV and $\mu_O - \mu_{O_2} \geq -3.35$ eV. Furthermore, Al_2O_3 is stabilized under N-rich and O-poor conditions satisfying $2(\mu_O - \mu_{O_2}) - (\mu_N - \mu_{N_{atom}}) \leq -3.84$ eV and $\mu_O - \mu_{O_2} \leq -3.35$ eV. The pressure of O_2 molecule at 1300 K corresponding to $\mu_O - \mu_{O_2} = -3.35$ eV is estimated to be ~ 1.0 Torr, so that even on Al_2O_3 ($1\bar{1}02$) surface the desorption of oxygen certainly occurs during the nitridation. We note that the surface with an N adatom (Al_2O_3–N_{ad}) shown in Fig. 10.14b is stabilized in a very narrow chemical potential range $(\mu_N - \mu_{N_{atom}} \geq -0.55$ eV). The surface with one substitutional N atom shown in Fig. 10.14c is always metastable over the entire chemical potential range.

Consequently, the remaining first and second layers in AlN–Al_2O_3 consist of two Al and two N atoms, respectively. Although in-plane positions of N atoms in AlN–Al_2O_3 are slightly different from those on $AlN(11\bar{2}0)$ surface (dashed line in Fig. 10.14e), the layer separation between first and second layers of AlN–Al_2O_3 (~ 0.4 Å) is close to that on $AlN(11\bar{2}0)$ surface (0.18 Å). Therefore, the oxygen desorption and nitrogen substitution can be regarded as a formation of one monolayer of nonpolar AlN layer whose in-plane alignment satisfies AlN[0001]‖ $Al_2O_3[1\bar{1}01]$. From this in-plane alignment, we can also deduce that the in-plane alignment between epitaxially grown nitrideson AlN layers and $Al_2O_3(1\bar{1}02)$ surface is [0001] nitrides $\|[1\bar{1}01]$ sapphire, consistent with the experimental results [65, 66].

10.4.3 Surface Phase Diagrams

In order to discuss the stability of nitrogen incorporated Al_2O_3 surfaces under realistic conditions, we furthermore estimate temperatures for structural change by comparing the adsorption energy of an NH_3 molecule with chemical potentials $\mu_{gas}(p, T)$ in (2.63). Since the calculated adsorption energies of H atoms for hydrogen terminated Al_2O_3 surfaces are found to be ranging from -0.07 to 0.61 eV per H atom, it is likely that the reactions for hydrogen adsorption are almost endothermic and surfaces including NH_3 fragments are metastable under the nitridation conditions. This implies that the decomposition of Al_2O_3 and AlN induced by hydrogen is not crucial at the initial stage of nitridation under low H_2 pressures [68]. We thus focus on the structural change between Al_2O_3–Al-I and AlN–Al_2O_3 on $Al_2O_3(0001)$ surface caused by NH_3 molecules. The substitution and desorption of O atoms by NH_3 molecules for Al_2O_3–I under NH_3 ambient conditions is considered through a surface reaction expressed as

$$Al_2O_3-Al-1+NH_3 \rightarrow AlN-Al_2O_3 + \frac{9}{2}H_2O + \frac{3}{4}O_2, \qquad (10.2)$$

and its reaction energy per NH_3 is compared with a net chemical potential $\mu_{NH_3} - (3/2)\mu_{H_2O} - (1/4)\mu_{O_2}$, where μ_{NH_3}, μ_{H_2O}, and μ_{O_2} are gas-phase chemical potentials of NH_3, H_2O, and O_2 molecules given by (2.63), respectively. The reaction in (10.2) is found to be endothermic and the reaction energy (per NH_3) is -5.82 eV. Figure 10.16a shows the calculated surface phase diagram on $Al_2O_3(0001)$ surface as functions of temperature and NH_3 pressure p_{NH_3} obtained by this comparison. Here, we assume low pressures $(1.0 \times 10^{-8}$ Torr) for μ_{H_2O} and μ_{O_2} since the amount of supplied NH_3 molecules is considered to be much larger than those of generated H_2O and O_2 molecules. The temperature for oxygen desorption leading to the formation of AlN–Al_2O_3 is 1330 K at $p_{NH_3} = 1.0 \times 10^{-8}$ Torr and decreases down to 605 K at $p_{NH_3} = 1.0 \times 10^{-2}$ Torr. This surface phase diagram clearly shows that oxygen atoms at the second layer desorb and N atoms substitute for the remaining O sites within the experimental temperature range $(1000 - 1300$ K) [46–53, 56–58, 62].

Similar temperature and pressure dependence to that on $Al_2O_3(0001)$ surface can be found on $Al_2O_3(1\bar{1}02)$ surface. In the case of $Al_2O_3(1\bar{1}02)$ surface, we can consider a surface reaction written as

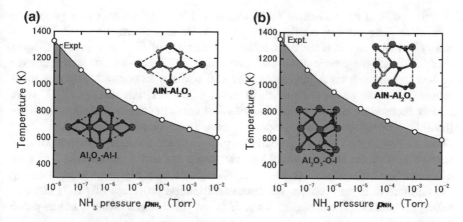

Fig. 10.16 Surface phase diagrams as functions of temperature and pressure of NH$_3$ molecule p_{NH_3} on **a** Al$_2$O$_3$(0001) and **b** Al$_2$O$_3$(1$\bar{1}$02) surfaces. We here assume low pressures $(1.0 \times 10^{-8}$ Torr) to calculate gas-phase chemical potentials of O$_2$ and H$_2$O. Top views of surface structures are also shown. Experimental temperature ranges for the nitridation on Al$_2$O$_3$(0001) [46–53, 56, 57, 61] and Al$_2$O$_3$(1$\bar{1}$02) [54, 55] substrates are attached in temperature axis. (Reproduced with permission from Akiyama et al. [63]. Copyright (2012) by Elsevier.)

$$Al_2O_3Al-I + 2NH_3 \rightarrow AlN-Al_2O_3 + 3H_2O + \frac{1}{2}O_2. \tag{10.3}$$

The reaction energy per NH$_3$ is compared with a net chemical potential $\mu_{NH_3} - (3/2)\mu_{H_2O} - (1/4)\mu_{O_2}$. The reaction in (10.3) is also endothermic with the reaction energy (per NH$_3$) of −5.78 eV. Figure 10.16b shows the calculated surface phase diagram on Al$_2$O$_3$(1$\bar{1}$02) surface as functions of temperature and p_{NH_3}. The temperature for oxygen desorption leading to the formation of AlN–Al$_2$O$_3$ is 1325 K at $p_{NH_3} = 1.0 \times 10^{-8}$ Torr and decreases down to 595 K at $p_{NH_3} = 1.0 \times 10^{-2}$ Torr. This surface phase diagram thus implies that oxygen atoms at the top layer desorb and N atoms substitute for those of third layer within the experimental temperature range (1300 − 1400 K) [54, 55]. Since the adsorption energy difference between Al$_2$O$_3$(0001) and (1$\bar{1}$02) surfaces for NH$_3$ molecules is small, there is little orientation dependence in the temperature range for AlN layer formation.

10.5 Chemical and Structural Change During Nitridation of Al$_2$O$_3$ Surfaces

In the preceding section, the stability of nitrogen incorporated Al$_2$O$_3$ surfaces has been systematically investigated on the basis of surface phase diagrams. The calculated surface energies demonstrate that the outermost O atoms of Al$_2$O$_3$ surfaces desorb and N atoms substitute for O atoms to form AlN layers under pregrowth conditions. The desorption of oxygen which leads to the formation of AlN layers has been found to be crucial for nitridation processes of Al_2O_3 substrates. Despite these findings, the atomic-scale understanding of chemical and structural change caused by nitrogen on Al_2O_3 surfaces is still unclear. In contrast, previous theoretical investigations on the surface in metal-oxide materials such as Al_2O_3 were mainly focused on surface stoichiometries and structures under realistic conditions [69–74]. The structures and properties have been found to change drastically depending upon the chemical potential of gas-phase species in equilibrium with the surface at different temperatures However, there are still open questions of how nitrided films with different stoichiometry compared with Al_2O_3 are formed during the nitridation. In this section, we discuss nitridation processes of Al_2O_3 surfaces taking account of realistic conditions such as temperature and pressure of gas-phase species [75].

Figure 10.17 shows the calculated surface energy under typical nitridation conditions ($T = 1400$ K, $p_N = 1 \times 10^{-4}$ Torr, and $p_{O_2} = 1 \times 10^{-8}$ Torr as a function of total number of atoms $N_{at} = \sum |n_i|$, which corresponds to the number of elemental processes during nitridation. Here, the configurations are labeled as $n_N/n_O/n_{Al}x$, where n_N, n_O, and n_{Al} are the numbers of additional N, O, and Al atoms with respect to the reference surface, respectively. As a reference surface, we use the stoichiometric Al-terminated and O-terminated surface structures for (0001) and ($1\bar{1}02$) orientations, respectively [71, 73]. The letter x specifies the different configurations with the same n_i, and the configurations are labeled simply as n_N/n_Ox when n_{Al} is zero. The stability of surfaces with different stoichiometries in equilibrium with the gas phase is determined by the surface energy in (4.1) as functions of chemical potentials. The surface energy $E_{surf}(n_N/n_O/n_{Al}x)$ at temperature T and pressure p_i of the ith type of atoms is given by

$$E_{surf}(n_N/n_O/n_{Al}x) = E_{tot}(n_N/n_O/n_{Al}x) - E_{ref} - \sum_i n_i \mu_i(p_i, T), \qquad (10.4)$$

where $E_{tot}(n_N/n_O/n_{Al}x)$ and E_{ref} are the total energy of the surface under consideration and of the reference surface. n_i is the number of excess ith-type atoms with respect to the reference and μ_i is the chemical potential at given temperature and pressure given by (2.63). Here, we assume that the chemical potential of Al is in equilibrium with bulk Al_2O_3. It is found that the surface energy decreases with the number of atoms and takes the lowest value for $N_{at} = 8$ and 5 on (0001) and ($1\bar{1}02$) orientations, respectively. The negative values of E_{surf} for large N_{at} indicate that structural changes toward the stable structures from the initial stoichiometric surfaces $[E_{surf}(0/0a) = 0$ eV] proceed under nitridation conditions.

For both orientations, these stable surfaces no longer maintain the stoichiometry of Al_2O_3. The inset of Fig. 10.17a shows the stable configuration, where three N atoms are substituted for O atoms and in total five O atoms desorb from the surface on (0001) orientation. For $Al_2O_3(1\bar{1}02)$ surface shown in Fig. 10.17b, two N atoms are substituted for O atoms and in total three O atoms desorb from the surface. If we

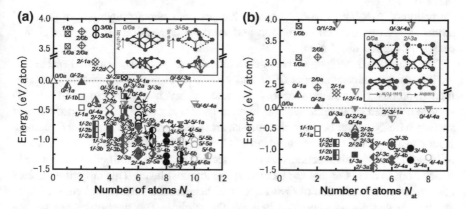

Fig. 10.17 Calculated surface energy of **a** $Al_2O_3(0001)$ and **b** $Al_2O_3(1\bar{1}02)$ surfaces as a function of total number of adsorption and desorption atoms $N_{at} = \sum |n_i|$. Only data for $-1.5 \leq E_{surf} \leq 4.0$ eV are displayed. Triangles correspond to the surfaces without N atoms, and squares, diamonds, circles, hexagons, and pentagons to the surfaces with one, two, three, four, and five nitrogen atoms in the unit cell, respectively. The surfaces with Al desorption are represented by inverted triangles. Patterns of symbols are defined by the difference between the number of incorporated N atoms and that of desorbing O atoms. Insets represent top and side views of stable surfaces with N atoms ($3/-5a$ and $2/-3a$ on $Al_2O_3(0001)$ and $Al_2O_3(1\bar{1}02)$ surfaces, respectively) and without N atom ($0/0a$), along with the in-plane directions. Large and filled (empty) small circles represent Al and O (N) atoms, respectively. (Reproduced with permission from Akiyama et al. [75]. Copyright (2013) by American Physical Society.)

take the stoichiometry of AlN and Al_2O_3 into account, the incorporation of two N atoms and desorption of three O atoms is at least required for AlN formation. These stable surfaces satisfy this condition and all the O atoms in the second anion layer are completely replaced by N atoms, so that they can be regarded as AlN layers generated by one monolayer (ML) of Al_2O_3. Furthermore, an important structural feature is also found for the interface between resultant AlN layer and the substrate. The insets of Fig. 10.17 show the stable structures $3/-5a$ and $3/-3a$ on (0001) and $(1\bar{1}02)$ orientations, which satisfy the in-plane alignment written as $AlN[1\bar{1}00] \| Al_2O_3[11\bar{2}0]$ and $AlN[0001] \| Al_2O_3[1\bar{1}01]$, respectively, as discussed in Sect. 10.4.

Moreover, another important feature of E_{surf} in Fig. 10.17 is related to the most stable configuration for each n_N/n_O with N atoms. The most stable configurations labeled as $n_N/n_O a$ in Fig. 10.17 contain substitutional N atoms beneath the surface. On the other hand, the energy of the surface with N atoms (labeled $n_N/n_O b$) at the topmost layer are 1.32 eV higher than that with N atoms beneath the surface. The surfaces with a substitutional N atom at the topmost layer are therefore metastable. The stabilization of a nitrogen incorporated surface is due to the formation of amphoteric (both covalent and ionic) Al–N bonds, whose bond energy is larger than that of Al–O bonds. The number of Al–N bonds in the substitutional N atom beneath the surface is larger than that in the N atom at the topmost layer. Therefore, the stabilization of N atoms beneath the surface is a driving force toward the formation of AlN layers, and structural changes such as rearrangement of O and N atoms as well as oxygen desorption and nitrogen adsorption are substantial processes during the nitridation.

The calculated structural change demonstrates several elemental processes leading to the rearrangements of O and N atoms. Figure 10.18 shows each elemental process, its reaction energy E_r, and energy barrier E_b on $Al_2O_3(0001)$ surface obtained by the nudged elastic band (NEB) method [76]. The formation of surface with N atoms beneath the surface from that with N atoms at the topmost layer consists of the outward diffusion of O atom followed by the inward diffusion of N atom. During this mechanism the surface becomes the intermediate structure in which both O and N atoms are located at the topmost layer. This structure appears after the outward diffusion of O atom located beneath the surface and consequently an oxygen vacancy is generated beneath the surface. The energy barrier for the diffusion is $0.91 - 2.83$ eV depending on the initial configuration. Owing to this oxygen vacancy the energy of intermediate structure is higher than that of the initial structure by $0.10 - 1.45$ eV. The surface transforms into the stable structure after the inward diffusion of N atom from the intermediate structure. The energy of transition state for N inward diffusion is higher than that of the intermediate state by $0.21 - 1.36$ eV. Since the energy of transition state for N indiffusion is lower than that for O outward diffusion, the energy barrier for a sequence of outward diffusion and inward diffusion is determined by the energy difference between the initial state and transition state for outward diffusion.

Similar results are obtained for geometries and energy profiles on $Al_2O_3(1\bar{1}02)$ surface shown in Fig. 10.19. After the outward diffusion of O atom the surface takes the intermediate structure and an oxygen vacancy is generated beneath the surface. The energy of intermediate structure is higher than that of the initial structure by $0.78 - 1.49$ eV. The surface transform into the stable structure after the inward diffusion of N atom from the intermediate structure. For the surface shown in Fig. 10.19a, however, the energy barrier is determined by the energy difference between the initial and transition states for nitrogen inward diffusion. The high energy of the transition state $(1/-2c^\wedge)$ in Fig. 10.19a can be attributed to the distance in diffusing N atom between $1/-2c$ and $1/-2a$. The N atom moves toward the surface by 2.73 Å during inward diffusion while the diffusion O atom does by 2.60 Å between $1/-2b$ and $1/-2c$ owing to this large distance for the N atom two Al–N bonds dissociate at the transition state.

Table 10.1 summarizes each elemental process, its reaction energy E_r, and energy barrier E_b obtained by the NEB method. The energy barrier for the concerted exchange [77] between the N atom located at the topmost layer and the O atom beneath the surface $(1/-1b \rightarrow 1/-1a)$ takes a quite large value [$E_b = 5.73$ and 5.70 eV on (0001) and $(1\bar{1}02)$ surfaces, respectively] compared with that of other elemental processes. The concerted exchange is thus virtually negligible and hardly involved in the incorporation of N atoms into the surfaces. In contrast, E_r takes a positive value and E_b is small for nitrogen inward diffusion for $\Delta n = n_O + n_N \leq 1$. This suggests that the inward diffusion of N atoms and the

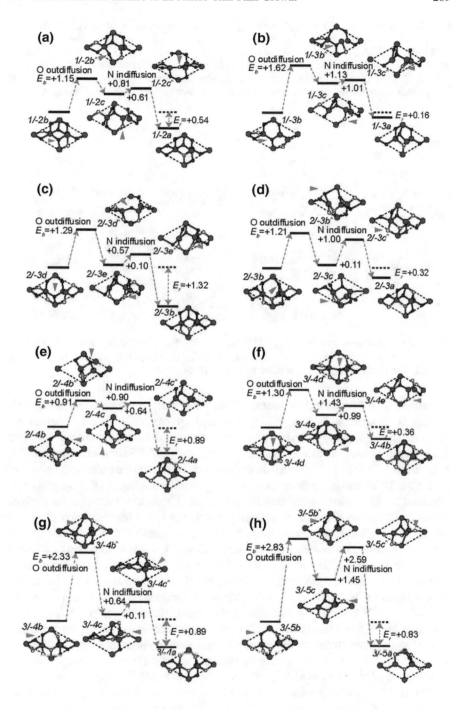

◄**Fig. 10.18** Geometries and energy profiles (in eV) for the diffusion of O and N atoms in
a $1/-2b \rightarrow 1/-2a$, **b** $1/-3b \rightarrow 1-3a$, **c** $2/-3e \rightarrow 2/-3b$, **d** $2/-3b \rightarrow 2/-3a$,
e $2/-4b \rightarrow 2/-4a$, **f** $3/-4d \rightarrow 3/-4b$, **g** $3/-4b \rightarrow 3/-4a$, and **h** $3/-5b \rightarrow 3/-5a$ on
$Al_2O_3(0001)$ surface obtained by the NEB method. The notations of atoms are same as in
Fig. 10.12. Only (meta) stable and transition state structures are depicted for simplicity. We add
"^" to each configuration for transition state structures. Arrowheads indicate diffusing N and O
atoms

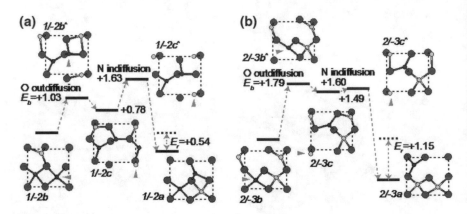

Fig. 10.19 Geometries and energy profiles (in eV) for the diffusion of O and N atoms in
a $1/-2b \rightarrow 1/-2a$ and **b** $2/-3b \rightarrow 2/-3a$ on $Al_2O_3(1\bar{1}02)$ surface obtained by the NEB
method. The notations of atoms are same as in Fig. 10.14

formation of substitutional N atoms beneath the surface occur when metastable
structures (labeled n_N/n_Oc or n_N/n_Oe) caused by oxygen outward diffusion are
formed. The high-energy barrier for oxygen outward diffusion compared with that
for nitrogen inward diffusion suggests that the outward diffusion of oxygen rather
than inward diffusion of nitrogen is expected to be a dominant elemental process.

Figure 10.20 shows geometries and energy profiles during nitridation processes
obtained by kMC simulations described in Sect. 2.3 on the basis of rates derived
from ab initio calculations. Although various types of elemental processes are
incorporated in the kMC simulations, the nitridation consists of a sequence of four
types of elemental processes. For both orientations, N atoms adsorb on the topmost
layer after the desorption of O atoms. The outward diffusion of oxygen located
beneath the surface and inward diffusion of the N atom (arrowheads in Fig. 10.20)
proceed after the nitrogen adsorption. This set of elemental processes repeats until 1
ML of an AlN layer is formed. Since the energy barriers for oxygen outward
diffusion are higher than those of nitrogen inward diffusion, the stable structures
with N atoms beneath the surface cannot be generated without the outward diffusion
of O atoms. It is thus concluded that, irrespective of surface orientation, outward
diffusion of O atoms located beneath the surface is a rate-limiting factor.

Figure 10.21 shows detailed geometries and energy profiles during nitridation on
$Al_2O_3(0001)$ surface obtained by kMC simulations on the basis of rates derived
from ab initio calculations. The desorption of O atoms and the adsorption of N atom

Table 10.1 Elemental process during structural change into AlN on Al_2O_3 surfaces. Initial and final states, reaction energy E_r (energy difference between initial and final states), energy barrier E_b obtained by the nudged elastic band method, and type of elemental process are listed. Positive (negative) values in E_r correspond to exothermic (endothermic) reactions. D_N, D_O, and C_{NO} denote inward diffusion of N atom, outward diffusion of O atom, and concerted exchange between N and O atoms, respectively. Geometries and energy profiles for inward diffusion of N atom and outward diffusion of O atom are depicted in Figs. 10.18 and 10.19

Orientation	Structures	E_r (eV)	E_b (eV)	Process type
(0001)	$1/-1b \rightarrow 1/-1a$	0.37	5.73	C_{NO}
	$1/-2b \rightarrow 1/-2c$	−0.61	1.15	D_O
	$1/-2c \rightarrow 1/-2a$	1.15	0.20	D_N
	$1/-3b \rightarrow 1/-3c$	−1.01	1.62	D_O
	$1/-3c \rightarrow 1/-3a$	1.17	0.12	D_N
	$2/-3d \rightarrow 2/-3e$	−0.10	1.29	D_O
	$2/-3e \rightarrow 2/-3b$	1.89	0.47	D_N
	$2/-3b \rightarrow 2/-3c$	−0.11	1.21	D_O
	$2/-3c \rightarrow 2/-3a$	0.43	0.89	D_N
	$2/-4b \rightarrow 2/-4c$	−0.90	0.91	D_O
	$2/-4c \rightarrow 2/-4a$	1.79	0.26	D_N
	$3/-4d \rightarrow 3/-4e$	−0.99	1.30	D_O
	$3/-4e \rightarrow 3/-4b$	1.35	0.44	D_N
	$3/-4b \rightarrow 3/-4c$	−0.11	2.33	D_O
	$3/-4c \rightarrow 3/-4a$	1.00	0.53	D_N
	$3/-5b \rightarrow 3/-5c$	−1.45	2.83	D_O
	$3/-5c \rightarrow 3/-5a$	2.28	1.14	D_N
($1\bar{1}02$)	$1/-1b \rightarrow 1/-1a$	0.23	5.70	C_{NO}
	$1/-2b \rightarrow 1/-2c$	−0.78	1.03	D_O
	$1/-2c \rightarrow 1/-2a$	1.32	0.85	D_N
	$2/-3b \rightarrow 2/-3c$	−1.49	1.79	D_O
	$2/-3c \rightarrow 2/-3a$	1.15	0.11	D_N

subsequently occur at the topmost layer. This is depicted from $0/0a$ to $1/-3b$ in Fig. 10.21. The outward diffusion of oxygen located beneath the surface and inward diffusion of the N atom (arrowheads in Fig. 10.21) proceed after the nitrogen adsorption leading to the formation of N-incorporated surface shown in $1/-3a$ of Fig. 10.21. This set of elemental processes can also be seen from $1/-3a$ to $2/-4a$ and from $2/-4a$ to $3/-5a$. As a result, 1 ML of AlN is formed by these structural changes. The analysis of kMC simulations clarifies that the most time-consuming process is oxygen outward diffusion from $3/-5b$ to $3/-5c$. It takes approximately 5×10^{-4} s to overcome this energy barrier at $T = 1400$ K. The relatively long duration of $3/-5b$ compared with those for other configuration is due to the energy barrier (2.83 eV) for oxygen diffusion.

Figure 10.22 shows detailed geometries and energy profiles during nitridation on $Al_2O_3(1\bar{1}02)$ surface obtained by kMC simulations. Similar to the case of

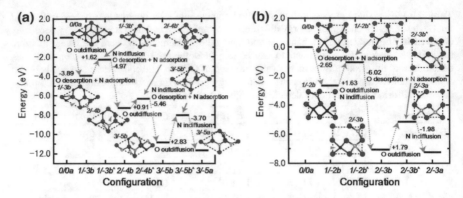

Fig. 10.20 Geometries and energy profiles (in eV) of **a** Al$_2$O$_3$(0001) and **b** Al$_2$O$_3$(1$\bar{1}$02) surfaces during nitridation processes obtained by kMC simularions under typical nitridation conditions ($T=$ 1400 K, $p_N = 1 \times 10^{-4}$ Torr, and $p_N = 1 \times 10^{-8}$ Torr). The notation of atoms are same as in Fig. 10.18. We add "^" to each configurations for transition state structures in O/N diffusion (for instance, $n_N/n_O^\wedge$). Arrowheads indicate diffusing N and O atoms. Note that the atomic coordinates in the figure are obtained by DFT calculations. (Reproduced with permission from Akiyama et al. [75]. Copyright (2013) by American Physical Society.)

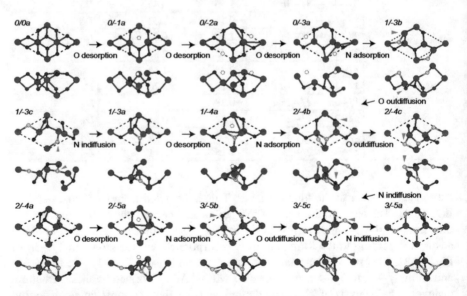

Fig. 10.21 Top views of typical configurations on Al$_2$O$_3$(0001) surface during nitridation obtained by kMC simulations at $T=$ 1400 K with $p_N = 1 \times 10^{-4}$ Torr, and $p_N = 1 \times 10^{-8}$ Torr. Filled and empty circles have the same meanings as in Fig. 10.18. Dashed circles represent oxygen vacancies caused by the desorption of O atoms at the topmost layer and arrowheads indicate diffusing N and O atoms. Note that the atomic coordinates in the figure are obtained by DFT calculations

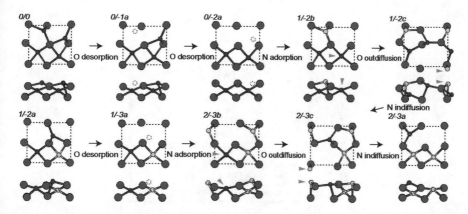

Fig. 10.22 Top views of typical configurations on $Al_2O_3(1\bar{1}02)$ surface during nitridation obtained by kMC simulations at $T = 1400$ K with $p_N = 1 \times 10^{-4}$ Torr, and $p_N = 1 \times 10^{-8}$ Torr. The notations are same as in Fig. 10.22

$Al_2O_3(0001)$ surface, the desorption of O atoms and the adsorption of N atom subsequently occur at the topmost layer, as depicted from $0/0a$ to $1/-2b$ in Fig. 10.22. The outward diffusion of oxygen and inward diffusion of the N atom take place from $1/-2b$ to $1/-2a$ and from $2/-3b$ to $2/-3a$, respectively. The most time consuming process is oxygen outward diffusion from $2/-3b$ to $2/-3c$, which is originating from its high energy barrier (1.79 eV). Owing to its low energy barrier compared to that from $3/-5b$ to $3/-5c$ on $Al_2O_3(0001)$ surface, however, the duration from $2/-3b$ to $2/-3c$ on $Al_2O_3(1\bar{1}02)$ surface ($\sim 1 \times 10^{-7}$ s) is shorter than that on $Al_2O_3(0001)$ surface.

The kMC simulations for various temperatures provide a consequence of oxygen diffusion. Figure 10.22 shows the growth rate for AlN formation as a function of reciprocal temperature estimated from kMC calculations. Here, the rate is defined as the reciprocal of time for AlN formation. The time for AlN formation is calculated by an integral of time increments until the formation of 1 ML of an AlN layer. Owing to an enhancement of oxygen diffusion at high temperatures, the rate increases monotonically with temperature. This trend is qualitatively consistent with the temperature dependence of XPS intensity corresponding to Al–N bonds [51].

Furthermore, several important features which are compared with the experiments can be deduced from the rate shown in Fig. 10.23. At each temperature the rate with high nitrogen pressure is larger than that with low pressure, indicating that the rate increases with nitrogen pressure. The pressure dependence results from high adsorption probability of N atoms under high nitrogen pressure conditions, and reasonably agrees with that in XPS intensity [51, 62]. Furthermore, the rate on $Al_2O_3(1\bar{1}02)$ surface is found to be always higher than that on an $Al_2O_3(0001)$ surface. The orientation dependence implies that the nitridation on nonpolar orientation is faster than that on polar (0001) orientation. This is because the energy

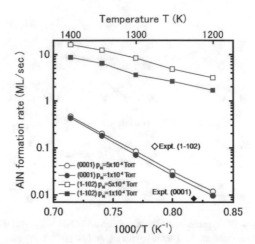

Fig. 10.23 Calculated nitridation rate on Al_2O_3 surfaces as a function of reciprocal temperature obtained by kMC simularions. Circles and squares represent the growth rate on $Al_2O_3(0001)$ and $(1\bar{1}02)$ surfaces, respectively. Filled and empty symbols denote the results with nitrogen pressures at $p_N = 1 \times 10^{-4}$ and 5×10^{-4} Torr, respectively. Experimental data for $Al_2O_3(0001)$ [58] and $(1\bar{1}02)$ [57] surfaces are shown as filled and empty diamonds, respectively. (Reproduced with permission from Akiyama et al. [75]. Copyright (2013) by American Physical Society.)

barrier for oxygen outward diffusion in $3/-5b$ on an $Al_2O_3(0001)$ surface (2.83 eV) is higher than that in $2/-3b$ on an $Al_2O_3(1\bar{1}02)$ surface (1.79 eV) due to the dissociation of two Al–O bonds. The overestimation of the calculated formation rate compared with the experiments might be due to the diffusion of N and O atoms in AlN layers with more than 2 MLs, which are not taken into account in the kMC simulations. Since these diffusion processes are expected to be insensitive to surface orientation, the formation rate is still dependent on oxygen outward diffusion, and the rate on nonpolar orientation could be faster than that on polar (0001) orientation even for AlN layers with more than 2 MLs. This trend is reasonably consistent with experimentally reported growth rate difference (diamonds in Fig. 10.23) [57, 58]. In order to predict the formation rate of AlN layers more quantitatively, clarifying further elemental processes such as diffusion of O and N atoms in AlN layers during nitridation should be clarified.

References

1. J. Wu, W. Walukiewicz, K.M. Yu, J.W. Ager III, E.E. Haller, H. Lu, W.J. Schaff, Y. Saito, Y. Nanishi, Unusual properties of the fundamental band gap of InN. Appl. Phys. Lett. **80**, 3967 (2002)
2. Y. Saito, H. Harima, E. Kurimoto, T. Yamaguchi, N. Teraguchi, A. Suzuki, T. Araki, Y. Nanishi, Growth temperature dependence of indium nitride crystalline quality grown by RF-MBE. Phys. Status Solidi B **234**, 796 (2002)

3. R.E. Jones, K.M. Yu, S.X. Li, W. Walukiewicz, J.W. Ager, E.E. Haller, H. Lu, W.J. Schaff, Evidence for p-type doping of InN. Phys. Rev. Lett. **96**, 125505 (2006)
4. P.A. Anderson, C.H. Swartz, D. Carder, R.J. Reeves, S.M. Durbina, S. Chandril, T.H. Myers, Buried p-type layers in Mg-doped InN. Appl. Phys. Lett. **89**, 184104 (2006)
5. V. Cimalla, M. Niebelschutz, G. Ecke, V. Lebedev, O. Ambacher, M. Himmerlich, S. Krischok, J.A. Schaefer, H. Lu, W.J. Schaff, Surface band bending at nominally undoped and Mg-doped InN by Auger electron spectroscopy. Phys. Status Solidi A **203**, 59 (2006)
6. X. Wang, S.-B. Che, Y. Ishitani, A. Yoshikawa, Growth and properties of Mg-doped In-polar InN films. Appl. Phys. Lett. **90**, 201913 (2007)
7. J.-H. Song, T. Akiyama, A.J. Freeman, Stabilization of bulk p-type and surface n-type carriers in Mg-doped InN{0001} films. Phys. Rev. Lett. **101**, 186801 (2008)
8. P. Waltereit, O. Brandt, A. Trampert, H.T. Grahn, J. Menniger, M. Ramsteiner, M. Reiche, K. H. Ploog, Nitride semiconductors free of electrostatic fields for efficient white light-emitting diodes. Nature **406**, 865 (2000)
9. M. Mclaurin, B. Haskell, S. Nakamura, J.S. Speck, Gallium adsorption onto ($11\bar{2}0$) gallium nitride surfaces. J. Appl. Phys. **96**, 327 (2004)
10. T. Takeuchi, S. Sota, M. Katsuragawa, M. Komori, H. Takeuchi, H. Amano, I. Akasaki, Quantum-confined stark effect due to piezoelectric fields in GaInN strained quantum wells. Jpn. J. Appl. Phys. **36**, L382 (1997)
11. P.D.C. King, T.D. Veal, C.F. McConville, F. Fuchs, J. Furthmuller, F. Bechstedt, P. Schley, R. Goldhahn, J. Schormann, D.J. As, K. Lischk, D. Muto, H. Naoi, Y. Nanishi, H. Lu, W. J. Schaff, Universality of electron accumulation at wurtzite c- and a-plane and zinc-blende InN surfaces. Appl. Phys. Lett. **91**, 092101 (2007)
12. C.-L. Wu, H.-M. Lee, C.-T. Kuo, C.-H. Chen, S. Gwo, Absence of fermi-level pinning at cleaved nonpolar InN surfaces. Phys. Rev. Lett. **101**, 106803 (2008)
13. D. Segev, C.G. van de Walle, Surface reconstructions on InN and GaN polar and nonpolar surfaces. Surf. Sci. **601**, L15 (2007)
14. C.G. Van de Walle, D. Segev, Microscopic origins of surface states on nitride surfaces. J. Appl. Phys. **101**, 081704 (2007)
15. C.K. Gan, D.J. Srolovitz, First-principles study of wurtzite InN(0001) and ($000\bar{1}$) surfaces. Phys. Rev. B **74**, 115319 (2006)
16. T.D. Veal, P.D.C. King, P.H. Jefferson, L.F.J. Piper, C.F. McConville, H. Lu, W.J. Schaff, P. A. Anderson, S.M. Durbin, D. Muto, H. Naoi, Y. Nanishi, In adlayers on c-plane InN surfaces: A polarity-dependent study by x-ray photoemission spectroscopy. Phys. Rev. B **76**, 075313 (2007)
17. J.E. Northrup, Effect of magnesium on the structure and growth of GaN(0001). Appl. Phys. Lett. **86**, 122108 (2005)
18. Q. Sun, A. Selloni, T.H. Myers, W.A. Doolittle, Energetics of Mg incorporation at GaN(0001) and GaN($000\bar{1}$) surfaces. Phys. Rev. B **73**, 155337 (2006)
19. M.D. Pashley, K.W. Haberern, W. Friday, J.M. Woodall, P.D. Kirchner, Structure of GaAs (001) (2×4)-c(2×8) determined by scanning tunneling microscopy. Phys. Rev. Lett. **60**, 2176 (1988)
20. T. Ito, K. Shiraishi, A theoretical investigation of migration potentials of Ga adatoms near kink and step edges on GaAs(001)-(2×4) surface. Jpn. J. Appl. Phys. **35**, L949 (1996)
21. T. Miyajima, S. Uemura, Y. Kudo, Y. Kitajima, A. Yamamoto, D. Muto, Y. Nanishi, GaN thin films on z- and x-cut LiNbO₃ substrates by MOVPE. Phys. Status Solidi C **5**, 1665 (2008)
22. D. Muto, H. Naoi, S. Takado, H. Na, T. Araki, Y. Nanishi, Mg-doped N-polar InN Grown by RF-MBE. Mater. Res. Soc. Symp. Proc. **955**, I08–01 (2007)
23. J.E. Northrup, Hydrogen and magnesium incorporation on c-plane and m-plane GaN surfaces. Phys. Rev. B **77**, 045313 (2008)

24. H. Amano, M. Kito, K. Hiramatsu, I. Akasaki, p-type conduction in Mg-doped GaN treated with low-energy electron beam irradiation (LEEBI). Jpn. J. Appl. Phys. **28**, L2112 (1989). Part 2
25. S. Nakamura, T. Mukai, M. Senoh, N. Iwasa, Thermal annealing effects on p-type Mg-doped GaN films. Jpn. J. Appl. Phys. **31**, L139 (1992). Part 2
26. C. Wang, R.F. Davis, Deposition of highly resistive, undoped, and p-type, magnesium-doped gallium nitride films by modified gas source molecular beam epitaxy. Appl. Phys. Lett. **63**, 990 (1993)
27. Z. Yang, L.K. Li, W.I. Wang, GaN grown by molecular beam epitaxy at high growth rates using ammonia as the nitrogen source. Appl. Phys. Lett. **67**, 1686 (1995)
28. F. Bernardini, V. Fiorentini, D. Vanderbilt, Spontaneous polarization and piezoelectric constants of III-V nitrides. Phys. Rev. B **56**, R10024 (1997)
29. T.J. Baker, B.A. Haskell, F. Wu, P.T. Fini, J.S. Speck, S. Nakamura, Characterization of planar semipolar gallium nitride films on spinel substrates. Jpn. J. Appl. Phys. **44**, L920 (2005). Part 2
30. M. Funato, M. Ueda, Y. Kawakami, Y. Narukawa, T. Kosugi, M. Takahashi, T. Mukai, Blue, green, and amber InGaN/GaN light-emitting diodes on semipolar {11-22} GaN bulk substrates. Jpn. J. Appl. Phys. **45**, L659 (2006). Part 2
31. M. Ueda, K. Kojima, M. Funato, Y. Kawakami, Y. Nakamura, T. Mukai, Epitaxial growth and optical properties of semipolar $(11\bar{2}2)$ GaN and InGaN/GaN quantum wells on GaN bulk substrates. Appl. Phys. Lett. **89**, 2101907 (2006)
32. J.F. Kaeding, H. Asamizu, H. Sato, M. Iza, T.E. Mates, S.P. DenBaars, J.S. Speck, S. Nakamura, Realization of high hole concentrations in Mg doped semipolar $(10\bar{1}1)$ GaN. Appl. Phys. Lett. **89**, 202104 (2006)
33. H. Zhong, A. Tyagi, N.N. Fellows, F. Wu, R.B. Chung, M. Saito, K. Fujito, J.S. Speck, S. P. DenBaars, S. Nakamura, High power and high efficiency blue light emitting diode on freestanding semipolar $(10\bar{1}\bar{1})$ bulk GaN substrate. Appl. Phys. Lett. **90**, 233504 (2007)
34. T. Hikosaka, N. Koide, Y. Honda, M. Yamaguchi, N. Sawaki, p-type conduction in a C-doped $(1\bar{1}01)$ GaN grown on a 7-degree-off oriented (001)Si substrate by selective MOVPE. Phys. Status Solidi C **3**, 1425 (2006)
35. J. Saida, E.H. Kim, T. Hikosaka, Y. Honda, M. Yamaguchi, N. Sawaki, Energy relaxation processes of photo-generated carriers in Mg doped (0001) GaN and $(1\bar{1}01)$ GaN. Phys. Status Solidi C **5**, 1746 (2008)
36. T. Hikosaka, N. Koide, Y. Honda, M. Yamaguchi, N. Sawaki, Mg doping in $(1\bar{1}01)$ GaN grown on a 7° off-axis (001)Si substrate by selective MOVPE. J. Cryst. Growth **298**, 207 (2007)
37. N. Sawaki, T. Hikosaka, N. Koide, S. Tanaka, Y. Honda, M. Yamaguchi, Growth and properties of semi-polar GaN on a patterned silicon substrate. J. Cryst. Growth **311**, 2867 (2009)
38. K. Tomita, T. Hikosaka, T. Kachi, N. Sawaki, Mg segregation in a $(1\bar{1}01)$ GaN grown on a 7° off-axis (001) Si substrate by MOVPE. J. Cryst. Growth **311**, 2883 (2009)
39. L.K. Li, M.J. Jurkovic, W.I. Wang, Surface polarity dependence of Mg doping in GaN grown by molecular-beam epitaxy. Appl. Phys. Lett. **76**, 1740 (2000)
40. E. Monroy, T. Andreev, P. Holliger, E. Bellet-Amalric, T. Shibata, M. Tanaka, B. Daudin, Modification of GaN(0001) growth kinetics by Mg doping. Appl. Phys. Lett. **84**, 2554 (2004)
41. T. Akiyama, D. Ammi, K. Nakamura, T. Ito, Stability of magnesium-Incorporated semipolar GaN$(10\bar{1}\bar{1})$ surfaces. Jpn. J. Appl. Phys. **48**, 110202 (2009)
42. T. Akiyama, D. Ammi, K. Nakamura, T. Ito, Surface reconstruction and magnesium incorporation on semipolar GaN surfaces. Phys. Rev. B **81**, 245317 (2010)
43. P. Boguslawski, E.L. Briggs, J. Bernholc, Amphoteric properties of substitutional carbon impurity in GaN and AlN. Appl. Phys. Lett. **69**, 233 (1996)
44. H. Tang, J.B. Webb, J.A. Bardwell, S. Raymond, J. Salzman, C. Uzan-Saguy, Properties of carbon-doped GaN. Appl. Phys. Lett. **78**, 757 (2001)

45. Y. Honda, N. Kameshiro, M. Yamaguchi, N. Sawaki, Growth of ($1\bar{1}01$) GaN on a 7-degree off-oriented (001)Si substrate by selective MOVPE. J. Cryst. Growth **242**, 82 (2002)
46. S. Keller, B.P. Keller, Y.F. Wu, B. Heying, D. Kapolnek, J.S. Speck, U.K. Mishra, S. P. DenBaars, Influence of sapphire nitridation on properties of gallium nitride grown by metalorganic chemical vapor deposition. Appl. Phys. Lett. **68**, 1525 (1996)
47. M.H. Kim, C. Sone, J.H. Yi, E. Yoon, Changes in the growth mode of low temperature GaN buffer layers with nitrogen plasma nitridation of sapphire substrates. Appl. Phys. Lett. **71**, 1228 (1997)
48. N. Grandjean, J. Massies, M. Leroux, Nitridation of sapphire. Effect on the optical properties of GaN epitaxial overlayers. Appl. Phys. Lett. **69**, 2071 (1996)
49. F. Widmann, G. Feuillet, B. Daudin, J.L. Rouviere, Low temperature sapphire nitridation: A clue to optimize GaN layers grown by molecular beam epitaxy. J. Appl. Phys. **85**, 1550 (1999)
50. K. Uchida, A. Watanabe, F. Yano, M. Kouguchi, T. Tanaka, S. Minagawa, Nitridation process of sapphire substrate surface and its effect on the growth of GaN. J. Appl. Phys. **79**, 3487 (1996)
51. T. Hashimoto, Y. Terakoshi, M. Ishida, M. Yuri, O. Imafuji, T. Sugino, A. Yoshikawa, K. Itoh, Structural investigation of sapphire surface after nitridation. J. Cryst. Growth **189–190**, 254 (1998)
52. M. Seelmann-Eggebert, H. Zimmermann, H. Obloh, R. Niebuhr, B. Wachtendorf, Plasma cleaning and nitridation of sapphire substrates for $Al_xGa_{1-x}N$ epitaxy as studied by x-ray photoelectron diffraction. J. Vac. Sci. Technol. **A16**, 2008 (1998)
53. P. Vennegues, B. Beaumont, Transmission electron microscopy study of the nitridation of the (0001) sapphire surface. Appl. Phys. Lett. **75**, 4115 (1999)
54. B. Ma, W. Hu, H. Miyake, K. Hiramatsu, Nitridating r-plane sapphire to improve crystal qualities and surface morphologies of a-plane GaN grown by metalorganic vapor phase epitaxy. Appl. Phys. Lett. **95**, 121910 (2009)
55. J.-J. Wu, Y. Katagiri, K. Okuura, D.-B. Li, H. Miyake, K. Hiramatsu, Effects of initial stages on the crystal quality of nonpolar a-plane AlN on r-plane sapphire by low-pressure HVPE. J. Cryst. Growth **311**, 3801 (2009)
56. N. Grandjean, J. Massies, Y. Martinez, P. Vennegues, M. Leroux, M. Laügt, GaN epitaxial growth on sapphire (0001): the role of the substrate nitridation. J. Cryst. Growth **178**, 220 (1997)
57. K. Masu, Y. Nakamura, T. Yamazaki, T.S.T. Tsubouchi, AlN/α-Al_2O_3 heteroepitaxial interface with initial-nitriding AlN layer. Jpn. J. Appl. Phys. **34**, L760 (1995)
58. M. Yeadon, M.T. Marshall, F. Hamdani, S. Pekin, H. Morkoç, J.M. Gibson, In situ transmission electron microscopy of AlN growth by nitridation of (0001) α-Al_2O_3. J. Appl. Phys. **83**, 2847 (1998)
59. M. Losurdo, P. Capezzuto, G. Bruno, Plasma cleaning and nitridation of sapphire (α-Al_2O_3) surfaces: new evidence from in situ real time ellipsometry. J. Appl. Phys. **88**, 2138 (2000)
60. G. Namkoong, W.A. Doolittle, A.S. Brown, M. Losurdo, P. Capezzuto, G. Bruno, Role of sapphire nitridation temperature on GaN growth by plasma assisted molecular beam epitaxy: Part I. Impact of the nitridation chemistry on material characteristics. J. Appl. Phys. **91**, 2499 (2002)
61. Y. Cho, Y. Kim, E.R. Weber, S. Ruvimov, Z. Liliental-Weber, Chemical and structural transformation of sapphire (Al2O3) surface by plasma source nitridation. J. Appl. Phys. **85**, 7909 (1999)
62. F. Dwikusuma, T.F. Kuech, X-ray photoelectron spectroscopic study on sapphire nitridation for GaN growth by hydride vapor phase epitaxy: nitridation mechanism. J. Appl. Phys. **94**, 5656 (2003)
63. T. Akiyama, Y. Saito, K. Nakamura, T. Ito, Stability of nitrogen incorporated Al_2O_3 surfaces: formation of AlN layers by oxygen desorption. Surf. Sci. **606**, 221 (2012)

64. T.M. French, G.A. Somorjai, Composition and surface structure of the (0001) face of alpha-alumina by low-energy electron diffraction. J. Phys. Chem. **74**, 2489 (1970)
65. M. Sano, M. Aoki, Epitaxial growth of undoped and Mg-doped GaN. Jpn. J. Appl. Phys. **15**, 1943 (1976)
66. T. Sasaki, S. Zembutsu, Substrate-orientation dependence of GaN single-crystal films grown by metalorganic vapor-phase epitaxy. J. Appl. Phys. **61**, 2533 (1987)
67. Th. Becker, A. Birkner, G. Witte, Ch. Wöll, Microstructure of the α-Al$_2$O$_3$(11$\bar{2}$0) surface. Phys. Rev. B **65**, 115401 (2002)
68. K. Akiyama, Y. Ishii, H. Murakami, Y. Kumagai, A. Koukitu, *In situ* gravimetric monitoring of surface reactions between sapphire and NH$_3$. J. Cryst. Growth **311**, 3110 (2009)
69. V. Puchin, J. Gale, A. Shluger, E. Kotomin, J. Günster, M. Brause, V. Kempter, Atomic and electronic structure of the corundum (0001) surface: comparison with surface spectroscopies. Surf. Sci. **370**, 190 (1997)
70. R. Di Felice, J.E. Northrup, Theory of the clean and hydrogenated Al$_2$O$_3$(0001)-(1×1) surfaces. Phys. Rev. B **60**, R16287 (1999)
71. X.-G. Wang, A. Chaka, M. Scheffler, Effect of the environment on α-Al$_2$O$_3$(0001) surface structures. Phys. Rev. Lett. **84**, 3650 (2000)
72. A. Marmier, S.C. Parker, ab initio morphology and surface thermodynamics of α−Al$_2$O$_3$. Phys. Rev. B **69**, 115409 (2004)
73. T. Kurita, K. Uchida, A. Oshiyama, Atomic and electronic structures of α-Al$_2$O$_3$ surfaces. Phys. Rev. B **82**, 155319 (2010)
74. J. Ahn, J.W. Rabalais, Composition and structure of the Al$_2$O$_3${0001}–(1×1) surface. Surf. Sci. **388**, 121 (1997)
75. T. Akiyama, Y. Saito, K. Nakamura, T. Ito, Nitridation of Al$_2$O$_3$ surfaces: chemical and structural change triggered by oxygen desorption. Phys. Rev. Lett. **110**, 026101 (2013)
76. G. Henkelman, B.P. Uberuaga, H. Jónsson, A climbing image nudged elastic band method for finding saddle points and minimum energy paths. J. Chem. Phys. **113**, 9901 (2000)
77. K.C. Pandey, Diffusion without vacancies or interstitials: a new concerted exchange mechanism. Phys. Rev. Lett. **57**, 2287 (1986)
78. T. Akiyama, K. Nakamura, T. Ito, J.-H. Song, A.J. Freeman, Structures and electronic states of Mg incorporated into InN surfaces: first-principles pseudopotential calculations. Phys. Rev. B **80**, 075316 (2009)
79. T. Akiyama, K. Nakamura, T. Ito, Stability of carbon incorpoated semipolar GaN(1$\bar{1}$01) surface. Jpn. J. Appl. Phys. **50**, 080216 (2011)

Index

Printed in the United States
By Bookmasters